GLYCOLIPIDS: NEW RESEARCH

GLYCOLIPIDS: NEW RESEARCH

DAIKI SASAKI
EDITOR

Nova Biomedical Books
New York

For permission to use material from this book please contact us:
Telephone 631-231-7269; Fax 631-231-8175
Web Site: http://www.novapublishers.com

Library of Congress Cataloging-in-Publication Data

Glycolipids : new research / Daiki Sasaki (editor).
p. ; cm.
ISBN 978-1-60456-216-3 (hardcover)
1. Glycolipids. I. Sasaki, Daiki.
[DNLM: 1. Glycolipids. QU 85 G5688 2007]
QP752.G56G59 2007
572'.57--dc22

2007047458

Published by Nova Science Publishers, Inc. ✦ New York

Contents

Preface

Glycolipids are carbohydrate-attached lipids. Their role is to provide energy and also serve as markers for cellular recognition. They occur where a carbohydrate chain is associated with phospholipids on the exoplasmic surface of the cell membrane. The carbohydrates are found on the outer surface of all eukaryotic cell membranes. They extend from the phospholipid bilayer into the aqueous environment outside the cell where it acts as a recognition site for specific chemicals as well as helping to maintain the stability of the membrane and attaching cells to one another to form tissues. This new book focuses on recent research results from around the world.

Short Communication A - Gaucher's disease, an inherited glycolipid storage disorder, is caused by a specific catabolic enzyme deficiency, this enzyme is called β-glucocerebrosidase (EC 3.2.1.45). Three clinical forms of Gaucher's disease have been described: type 1, or non-neuropathic; type 2, or acute neuropathic; and type 3, or subacute neuropathic.

Since 1980, scientists have had to answer several requests over the diagnosis of this metabolic disease. Requests emanating primarily from paediatrics departments. Fifteen cases were confirmed, within sight the intraleucocytic β-glucocerebrosidase enzyme activity measurement.

The authors report their experiment in the biochemical diagnosis of Gaucher disease by showing mainly the variability and the extreme heterogeneity of beta glucocerebrosidase activity during performed measurements. In addition, the authors expose the problems of etiologic diagnosis which certain patients raise, given the discordances between measured enzymatic activity and clinical signs of the disease in a person with Gaucher's disease. Moreover, the authors develop the biological parameters useful to proportion to the monitoring of the treatment, finally available.

Short Communication B - It is known that the concentration of sulfatide (SM4s), a major sulfoglycolipid in Madin-Darby canine kidney (MDCK) cells increases in a hyperosmotic medium due to the activation of the cerebroside sulfotransferase gene. The authors investigated the changes in molecular species of sulfoglycolipids in this cell line under osmotic stresses. Analysis by thin layer chromatography showed that the slower-moving band of SM4s specifically increased in hyperosmotic medium. Moreover, analysis by negative ion liquid secondary ion mass spectrometry demonstrated an increase in the ratios of hydroxy fatty acid species to those of corresponding non-hydroxy fatty acid ones of SM4s, under hyperosmotic stress. Further, analysis of galactosylceramide, the precursor of SM4s, by negative ion LSIMS also showed a similar result, which suggests that the hydroxylation of fatty acids in sulfatide was specifically induced through the hydroxylation of its precursor glycolipid, under hyperosmotic stress. The elucidation of the mechanism that increases the hydrophilic molecular species in galactosphingolipids of renal cells under hyperosmotic stress may contribute to the *in vitro* study on the function of galactolipids containing hydroxy fatty acids.

Chapter I - Glycolipids—in particular gangliosides—are ubiquitous molecules on the surface of cells. They play roles in important cellular processes like differentiation and transformation. This chapter focuses on the role of gangliosides in the interaction of eukaryotic cells with microbes and microbe-derived elements, such as toxins, as well as the role of gangliosides in the activation of immune cells. Gangliosides can be considered cellular gateways for certain microbes and viruses that have the ability to interact and to invade eukaryotic cells. Most of these early interactions during infection initially take place in the lipid raft domains of the plasma membrane. Gangliosides also regulate the immune system and play a role in the efficacy of certain experimental mucosal adjuvants (i.e., substances that enhance the immune response against antigens). However, their mechanism of action is not yet well elucidated. Consequently, the outcome of some host-pathogen interactions may be manipulated at two distinct levels: at the initial host-pathogen interaction and through intervention at the specific immune response that is induced later.

Moreover, gangliosides serve as receptors for some host molecules, such as antibodies. In addition, their attachment can lead to the recruitment and activation of toll-like receptors and signaling transduction pathways that promote interactions and activate diverse type of cells. This chapter will explore the participation of gangliosides and lipid rafts in these diverse events. A better understanding of the molecular mechanism involved in these interactions will allow us to manipulate host-pathogen and host immune responses more effectively to prevent infectious diseases and to optimize vaccination and prophylaxis against certain diseases.

Chapter II - Gangliosides are complex glycosphingolipids present in all eukaryotic cells. They are composed of a hydrophobic ceramide backbone anchored in the lipid bilayer of plasma membrane and a polar head group with hydrophobic oligosaccharide chains and one or more residues of *N*-acetylneuraminic acid (sialic acid) at the termini that protrude into the extracellular space. Ganglioside species vary one from another in their ceramide composition and/or in their carbohydrate content. The kinds and amounts of gangliosides expressed in the plasma membrane are cell and tissue type dependent. Ganglioside composition may change during embryonic development, tissue differentiation, aging and degeneration or malignant transformation. Gangliosides have been recognized to play important roles in normal physiological functions as well as in pathological conditions, including tumor progression. They modulate many biologic processes, such as differentiation, proliferation, adhesion, migration, immune response, and act as receptors for viruses and bacterial toxins. Gangliosides also participate in apoptosis and signal transduction in several cell types. In mast cells, gangliosides may function in modulating secretory events as well as participating in mast cell development and recruitment. Mast cells have long been known to play a critical role in inflammatory and allergic reactions and contribute to host resistance. Following mast cell activation by cross-linking of the high affinity IgE surface receptor (FcεRI) on the cell surface a number of inflammatory mediators are released. Previous studies using the monoclonal antibody (mAb) AA4, which recognizes two α-galactosyl derivatives of ganglioside GD_{1b} that are specific to rodent mast cells showed that these gangliosides are located in proximity to FcεRI. Binding of mAb AA4 to the gangliosides on the surface of mast cells produces morphological and biochemical changes similar to those seen with activation of FcεRI, but without histamine release. The expression of these gangliosides on the mast cell surface has also been shown to be related to mast cell maturation. Committed mast cell precursors do not express the gangliosides recognized by mAb AA4, but the gangliosides begin to be expressed on the cell surface jointly with FcεRI in very immature mast cells and continue to be expressed by mast cells in all stages of maturation. This chapter will present a short review on the biological functions of gangliosides and discuss some important evidence on the role of gangliosides in modulating FcεRI and in the maintenance of mast cell structure and function.

Chapter III - Glycolipids are, by definition, long carbohydrate and hydrophobic molecules that integrate into the plasma membrane and bind to diverse types of sugar residues at their extracellular domain. Their location on the external surface of the cell makes them excellent biomarkers for the prospective isolation of various subsets of cells from the central nervous system. In this chapter, the authors review glycolipids that have been used to identify diverse populations of neural stem and progenitor cells during development and in the adult. In particular, the authors focus our discussion on *stage specific embryonic antigen 1* and *4* (SSEA-1 and SSEA-4, respectively). These two glycolipids are widely used to monitor embryonic stem cell differentiation, but have more recently been employed in the identification and positive selection of two distinct subsets of neural stem and progenitor cells. The authors explore their possible biological functions and discuss differences between species. And finally, the authors speculate on their possible applications in the therapeutic correction of neurological damage.

Chapter IV - Mycobacterial diseases, including tuberculosis, leprosy, and disease due to nontuberculous mycobacteria, are the major cause of death from infectious diseases around the world. About one-third of the world population is latently infected with *Mycobacterium tuberculosis*. Over 8 million new cases and nearly 2 million deaths occur each year resulted from tuberculosis. Tuberculosis presents a significant health threat to the world. The pathogenicity of mycobacteria is related to their ability to escape killing by ingested macrophages, latent infection, and induce delayed-type hypersensitivity, such as granulomas. This has been attributed to several components of the mycobacterial cell wall, such as, complement activation factor, heat-shock protein, mycobacterial DNA binding protein, and glycolipids, including mycoloyl glycolipids (containing cord factor/trehalose dimycolate), phenolic glycolipid, sulfolipid, glycopeptidolipid, phosphatidylinositol mannoside, lipoarabinomannan, and lipomannan. Among them, glycolipids specific for mycobacteria play a key role in the pathogenesis, because mycobacteria are rich in lipids that possess pleiotropic activities. The complex interaction between a range of mycobacterial components and the host participate the pathogenesis. In this review the authors will focus on the pleiotropic activity of glycolipids in both the microbe and host. The better understanding of mycobacterial pathogenicity may open the new avenue for the development of diagnostics, therapeutic and prophylactic interventions.

Chapter V - In autoimmune neuropathies such as Guillain-Barré syndrome (GBS) and IgM paraproteinemic neuropathy, serum antiglycolipid antibodies are frequently elevated in titer. Those antibodies are useful diagnostic markers and some of them may be directly involved in the pathogenetic mechanisms. In GBS, the carbohydrate structures of the microorganisms of the antecedent infections that mimic those of human glycolipids may induce production of the antiglycolipid antibodies. "Molecular mimicry" mechanism has been shown in the production of antiganglioside antibodies after *Campylobacter jejuni* infection and antigalactocerebroside antibodies after *Mycoplasma pneumoniae* infection. The antibodies may specifically bind to the regions where the respective antigens are located (for example, the antibodies against the ganglioside GQ1b may cause ophthalmoplegia by binding to the paranodal regions of the oculomotor, trochlear, and abducens nerves, where GQ1b is densely localized). The authors have proved this hypothesis by inducing sensory ataxic neuropathy in rabbits sensitized with the ganglioside GD1b, which is localized in the primary sensory neurons conveying deep sensation. The authors have recently found that the antibodies that specifically recognize a new conformational epitope formed by two gangliosides are sometimes present in the acute-phase sera of GBS patients. In particular, the antibodies against GD1a-GD1b complex are associated with severe GBS requiring artificial ventilation. Gangliosides are preferentially packaged with cholesterol to form lipid rafts, in which the carbohydrate portions of two different gangliosides may form a new conformational epitope. Within the rafts, gangliosides are considered to interact with important receptors or signal transducers. The antibodies against ganglioside complex may therefore directly induce nerve conduction failure and severe disability in GBS. More study is needed to elucidate the roles of antiglycolipid antibodies in the pathogenetic mechanisms of autoimmune neuropathies.

Chapter VI - The membrane raft hypothesis proposed that membrane lipid domains enriched in Chol and sphingolipids form dynamic sorting and signalling platforms for membrane proteins in cells. Cell membranes however cannot be accurately described with only the two categories of raft or non-raft; instead the authors should talk about different domains which should be characterized each in all its complexities. In this chapter the current picture on membrane domains is discussed, and the usefulness of model systems to understand molecular configuration is exemplified by monolayer isotherms, calorimetry and spectroscopy studies in model membranes containing ceramide. These results are interpreted with the aim of explaining the ability of ceramide to produce very large domains or "platforms" in cell membranes.

Chapter VII - Ceramide is an apoptosis-inducing signaling molecule in the sphingolipid metabolism which emerges as a potential target for anticancer therapy. Natural ceramide in solution has a uniquely folded butterfly conformer with strong internal hydrogen bonding between amide and two alcohols at C1 and C3 and with hydrophobic interaction between two long alkyl chain of sphingosine backbone and fatty acid. Its heterocyclic analogs as 3-alkanoyl or benzoyl-4-(1-hydroxyhexadec-2-enyl)-oxazolidin-2-ones were designed by binding of primary alcohol and amide in sphinogosine backbone as a carbamate. Those compounds without free primary alcohol have advantages not to perturb the ceramide metabolism most of which involved in the primary alcohol of sphingosine backbone. This structurally constrained analog of ceramide may have better activity compared with the natural ceramide without pre-organization of substrate, if necessary. They were synthesized by addition of acyl halide to the common ring, 4-(1-*t*-butyldimethylsilyloxyhexadec-2-enyl)-oxazolidin-2-one, which was elaborated from chiral aziridine-2-carboxylate including stereoselective reduction and ring opening reactions. All compounds were tested as antileukemic drugs against human leukemia HL-60 cells. Many of them including propionyl, cyclopentanoyl and *p*-nitrobenzoyl-4-(1-hydroxyhexadec-2-enyl)-oxazolidin-2-ones showed better antileukemic activities than natural C2-ceramide with good correlation between cell death and DNA fragmentation. Cytotoxicity was induced by caspase activation without significant accumulation of endogenous ceramide concentration or any perturbation of ceramide metabolism.

Chapter VIII - Increasing lines of evidence suggest that cholesterol is a potent risk factor for the development of Alzheimer's disease (AD). An increase in the cholesterol level of neuronal membranes likely enhances both the generation and aggregation of the amyloid β-protein (Aβ). Regarding the enhancement of Aß aggregation, previous studies by the authors group and other groups suggest that cholesterol has both direct and indirect effects on such aggregation. The authors previously identified a novel Aβ species in the brain exhibiting early pathological changes of AD. On the basis of unique molecular characteristics, the authors hypothesized that GM1 ganglioside (GM1) plays a critical role in the pathological processes in AD. The mechanism underlying the binding of Aβ to GM1 ganglioside remains to be determined; however, their recent studies suggest that an increase in the cholesterol level in the membrane facilitates such binding. The distribution of cholesterol in neuronal membranes is likely to be altered by aging and the expression of apolipoprotein E4, both of which are strong risk factors for AD development. The putative roles of cholesterol and gangliosides in AD development are discussed.

Chapter IX - Basidiomycetous yeasts *Cryptococcus humicola* and *Pseudozyma fusiformata* secrete into the culture broth glycolipids which are a 16-(β-cellobiosyloxy)-hexadecanoic acid containing OH-groups at C-2 or C-2 and C-15 of fatty acid aglycon and also carrying O-acyl substituents in cellobiosyl fragment. In the case of *Cr. humicola,* 16-(tetra-O-acetyl-β-cellobiosyloxy)-2,15-dihydrohexadecanoic acid (1) and 16-(tetra-O-acetyl-β-cellobiosyloxy)-2-hydroxyhexadecanoic acid (2) were structurally defined as major products, while *P. fusiformata* secreted mainly 16-[6-O-acetyl-2'-O-(3-hydroxyhexanoyl)-β-cellobiosyloxy)-2,15-dihydroxyhexadecanoic acid (3). These compounds are lethal for many yeast and mycelial fungi. They exhibit higher activity at pH ~4.0 and in the temperature range of 20-30 °C, and had similar fungicidal activities against different yeasts including pathogenic *Cryptococcus* and *Candida* species. The cells of tremellaceous yeast *Filobasidiella neoformans* causing systemic cryptococcosis completely died after 30-min incubation with glycolipids 1-3 at 0.02 mg/ml concentrations. The same effect on ascomycetous yeast, including pathogenic *Candida* species, is achieved at five to eight times higher concentrations of glycolipids. Cellobiose lipid 3 of *P. fusiformata* inhibits the growth of phytopathogenic fungi *Sclerotinia sclerotiorum* and *Phomopsis helianthi* more efficiently than cellobiose lipids **1** and **2** from *Cr. humicola.*

Fully O-deacylated analogues of compounds 1-3, namely 16-(β-cellobiosyloxy)-2,15-dihydroxyhexadecanoic acid (4) and 16-(β-cellobiosyloxy)-2-hydroxyhexadecanoic acid (5), were studied together with a totally synthetic compound, 16-(β-cellobiosyloxy)-hexadecanoic acid (6), to assess the structural features determining their fungicidal activity. Compound 6 without OH-group and compound 5 with only one OH-group in the aglycon do not inhibit the growth of *F. neoformans* and *Saccharomyces cerevisiae*, while compound 4 double-hydroxylated in the fatty acid moiety inhibits the growth of both test cultures but at 5-10 times higher concentrations than compounds 1 and 2 from *Cr. humicola.* Thus, the structures of both the carbohydrate part and fatty acid aglycon moiety influence the fungicidal activity of cellobiose lipids under study. The treatment with cellobiose lipids causes the leakage of ATP and potassium ions from yeast cells. It suggests that the fungicidal activity of cellobiose lipids is due to their ability to enhance the nonspecific permeability of the cytoplasmic membrane. The effective concentrations of cellobiose lipids, which induce ATP release and cell death, were similar for the same test cultures. The results show that extracellular cellobiose lipids may be considered as promising fungicides for agriculture, medicine, and veterinary.

In: Glycolipids: New Research
Editor: Daiki Sasaki, pp. 1-18
ISBN: 978-1-60456-216-3

Short Communication A

The Biochemical Diagnosis of Gaucher Disease

Lyece Yargui*
Medical Biochemistry
Faculty of Medicine - Algiers University

Abstract

Gaucher's disease, an inherited glycolipid storage disorder, is caused by a specific catabolic enzyme deficiency, this enzyme is called β-glucocerebrosidase (EC 3.2.1.45). Three clinical forms of Gaucher's disease have been described: type 1, or non-neuropathic; type 2, or acute neuropathic; and type 3, or subacute neuropathic.

Since 1980, scientists have had to answer several requests over the diagnosis of this metabolic disease. Requests emanating primarily from paediatrics departments. Fifteen cases were confirmed, within sight the intraleucocytic β-glucocerebrosidase enzyme activity measurement.

I, hereby, report my experiment in the biochemical diagnosis of Gaucher disease by showing mainly the variability and the extreme heterogeneity of beta glucocerebrosidase activity during performed measurements. In addition, I expose the problems of etiologic diagnosis which certain patients raise, given the discordances between measured enzymatic activity and clinical signs of the disease in a person with Gaucher's disease. Moreover, I develop the biological parameters useful to proportion to the monitoring of the treatment, finally available.

* Professional address: Laboratoire Central de Biochimie – CHU Mustapha Alger. Place 1er mai. 16000 ALGIERS. ALGERIA. Home address : 04 rue Fatima. 06000 BEJAIA. ALGERIA. Cell phone (direct) Tel : 00 213 50 178 111. E-mail : lyecebiochem@yahoo.fr

Introduction

Gaucher's disease is a lysosomal overload disease, that we often, encountered during the last twenty years. Indeed, since 1980, we (my team and I) have had to answer several requests over the diagnosis of this metabolic disease, requests emanating primarily from paediatrics departments. Fifteen cases were confirmed, based on the intraleucocytic activity of β-glucocerebrosidase, an enzyme involved in sphingolipids catabolism.

This intraleucocytic measurement is far more significant than the histological test that existed so far, and needed characteristic bone marrow storage cells.

Gaucher's disease importance has risen with the arrival of a substitutive treatment, currently available of the market.

Therefore, metabolic diseases are no more an irreversible fatality as, plenty of innate metabolic errors benefit by a treatment.

Some of these diseases can be curable; in particular with a dietetic treatment eliminating a compound that the organism can not use (PKU), or with the in-take of a vitaminic co-factor that can not be synthesized (homocystinuria, certain types or ataxia and PKU). The use of inhibiting substrates (Gaucher, Tay-Sacchs, tryrosinémia type I, Leish nyhan), of chelator subtrates (friedreich ataxia), or the replacement of the failing enzyme (Gaucher, Fabry, Hurler, Maroteaux Lamy) are, aswell efficient therapies (Table 1).

Table 1. Treatment of Metabolique Diseases

THERAPEUTIQUE	METABOLIQUE DISEASE	SPECIALITY
Dietetic treatment	Phenylkenoturia	*LOPHENALAC* (MEAD JOHNSON)
Vitamin therapy	Homocystinuria type I	*VITAMINS* B_6-B_9-B_{12}
	Phenylkenoturia (not all types)	*VITAMIN* BH_4
	Ataxia (not all types)	*VITAMIN E*
Substitutives hormones	Hypothyroidia	*LEVOTHYROX* (MERCK LIPHA SANTE)
Replacement of the defect transporter	Cystinosis	*CYSTEAMINE (CYSTAGON®)* (ORPHAN PHARMACEUTICALS)
Replacement of the defect enzyme	Gaucher	*CEREZYME* (GENZYME)
	Fabry	*FABRAZYME* (GENZYME) *REPLAGAL* (TKT)
	Hurler	*ALDURAZYME* (GENZYME) (BIOMARIN)
	Maroteaux - Lamy	*NAGLAZIME* (GENZYME) (BIOMARIN)

THERAPEUTIQUE	METABOLIQUE DISEASE	SPECIALITY
Inhibitors of enzymes	Gaucher Tay-Sachs	*OGT 918 or ZAVESCA* (ACTELION LTD)
	Tyrosinemia Type I	*NTBC* 2-(2-nitro-4-trifluoromethylbenzoyl)-1,3 cyclohexanedione (ORPHAN PHARMACEUTICALS)
	Lesch-Nyhan	*ALLOPURINOL* (GSK – GLASKO SMITH KLINE)
Potent free-radical scavenger	Friedreich's ataxia	*IDEBENONE* (TAKEDA PHARMA, JAPAN)

The objective of this work is to provide the essential elements to the biochemical diagnosis of Gaucher's disease supported by the experience I acquired from my elders.

Before getting to develop this laboratory practice, a brief summary on sphingolipids metabolism seem to be necessary in order to clarify the complexity often, encountered to define the enzymatic bloc causing metabolic diseases.

I. Brief Summary on Sphingolipids Metabolism

I.1. Sphingolipids Structure

Sphingolipids are complex lipid molecules. It constitutes plasmic membranes of white blood cells and erythrocytes.

Their structure (Figure 1) has in commun a ceramid that is constitued of an alcohol called sphingosine linked to a long-chain fatty acid(palmitate C18 : 0, stearate C16 : 0, or more rarely nevronate C24 : 1 or behenoate C22: 0).

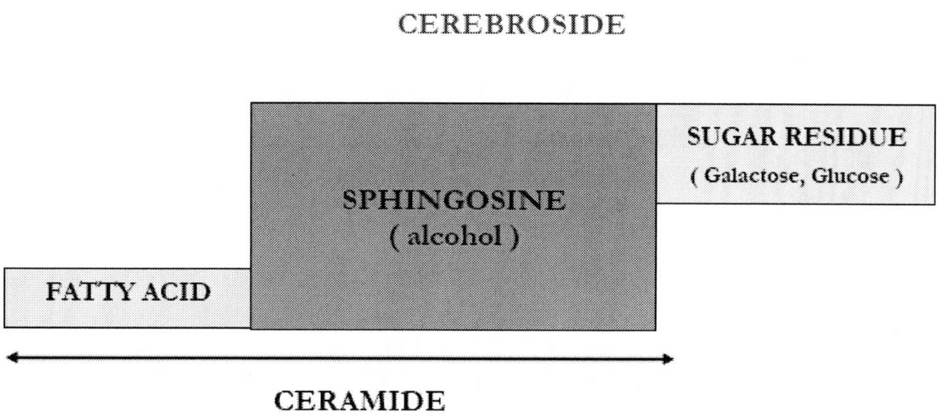

Figure 1. (Continued).

GANGLIOSIDE (example of G_{M1})

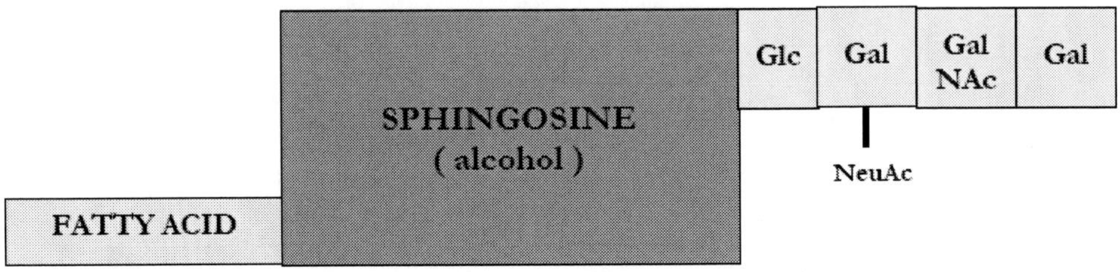

G_{M1} : G,Ganglioside ; M, Monosialo ; 3, number assigned on the basis of the chromatographic migration
Glc : Glucose ; Gal : Galactose ; NeuAc : N-acetylneuraminic acid ;
Gal Nac : N- acetyl - Galactosamine

SPHINGOMYELIN

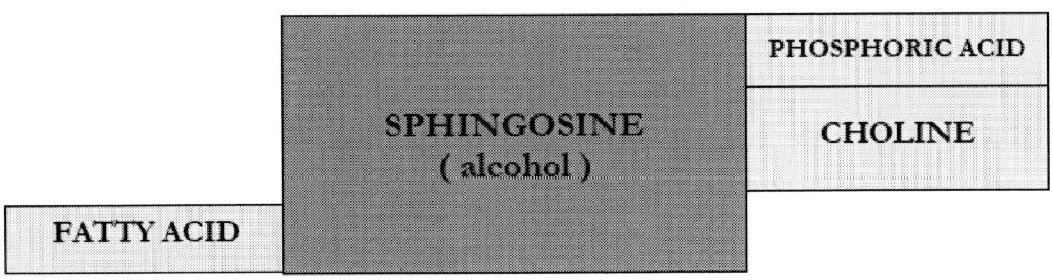

Figure 1. Sphingolipid's structure.

It is the fraction added to this ceramid that will make the difference between the several existing sphingolipids. And it is this fraction that is the origin of most metabolic abnormalities. This way, we will talk:

- about cerebrosides when we associate a ceramid to a simple sugar that can be glucose or galactose;
- About gangliosides when several sugars or sugar derivates (sialic acid) are associated to the ceramid;
- About sphingomyeline when a phosphorylcholine is linked to a ceramid.

I.2. Sphingolipidosis

Enzymatic deficiencies affecting the catabolic path of sphingolipids are gathered under the name of sphingolipidosis

Several enzymatic blocs mark out this degradation path (Figure 2 and 3):

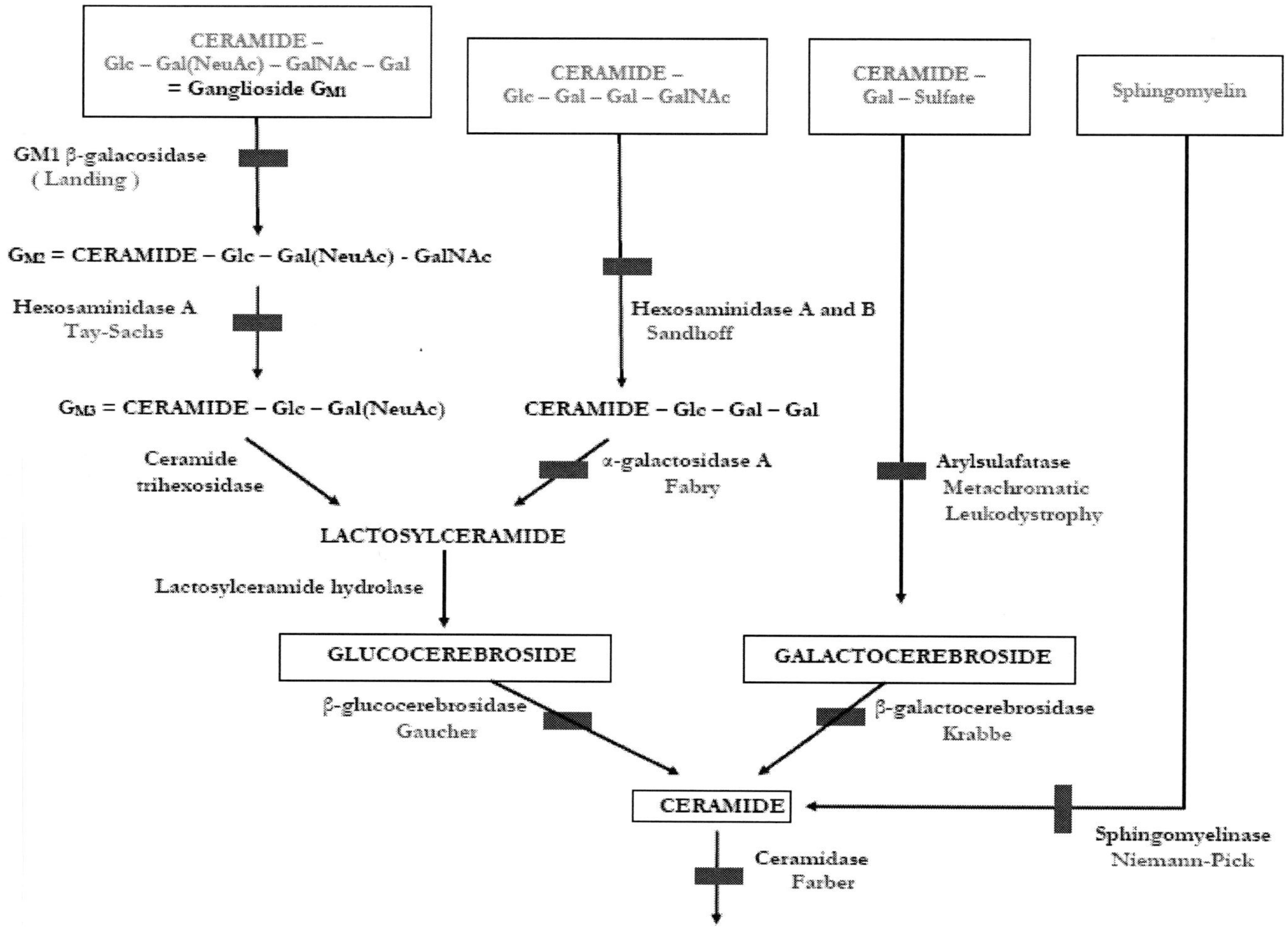

Figure 2. The enzymatic defect in the lysosomal degradation pathway of sphingolipides.

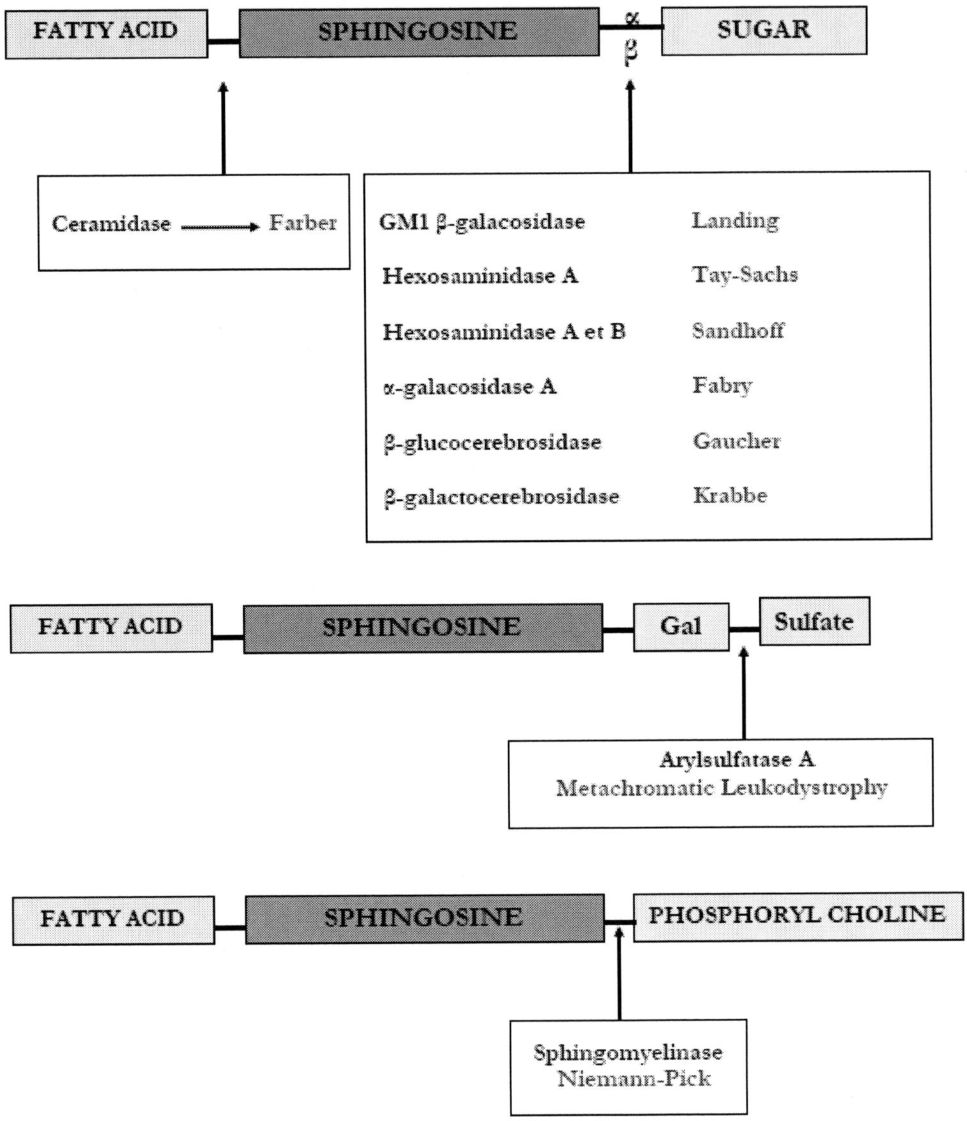

Figure 3. Sites of deficient enzyme reactions in sphingolipidoses (Simplified diagram).

a) Those concerning enzymes linking in the position Alpha or beta the sugar to the ceramid: Therefore, we distinguish:

- Landing disease with the deficiency of GM1 β-galacosidase;
- Tay-Sachs with the deficiency of Hexosaminidase A;
- Sandhoff disease with the deficiency of Hexosaminidase A and B;
- Fabry disease with the deficiency of α-galacosidase A ;
- Gaucher disease with the deficiency of β-glucocerebrosidase ;
- Krabb disease with the deficiency of β-galactocerébrosidase.

b) Arylsulfatase A deficiency, the enzyme that hydrolyses the bond between the sulfate group and the galactoceramid. The disease caused is Metachromatic Leucodystrophia.
c) The deficiency of sphingomyelinase, enzyme which allows to separate phosphocholine of ceramid. This is Nieman Pick disease.
d) Finally, the enzymatic bloc may concern the enzyme separating the fatty acid of sphingosine. This is another metabolic disease: Farber lipogranulomatosis.

The central common thing of all of these sphingolipidosis is the accumulation of metabolites which are not degraded in the cell,because of lysosomal digestion defect. This will trigger a cascade of metabolic abnormalities affecting cells of vital function organs: liver, spleen, marrow bone and nervous system [1,2].

I.3. Gaucher's Disease

The molecule that the metabolic abnormality in relation with Gaucher's disease concerns glucocerebroside called D-glucosylceramid. This is a ceramid linked in β position with the glucose coming from the degradation of a gangliosid, an even more complex molecule (Figure 4).

BRADY [3] and PATRICK [4] were the first to demonstrate that the accumulation of glucucocerebroside in the reticuloendothelial system of patients with Gaucher's disease (GD) is the result of a deficiency in an enzyme, which catalyses the cleavage of glucose from glucocerebroside. This enzymatic defect has since been found by various investigators in the spleen of patients as well as in other tissues, including skin fibroblasts and white blood cells [5]. The enzymatic deficiency in question affects the β-glucocerebrosidase = β- acid glucosidase = Glucosylceramide hydrolase: an enzyme (EC 3.2.1.21) that hydrolyses the link of the bond β between ceramid and D-glucose.

This enzyme has some quite particular physicochemical characteristics. Indeed, it acts at various pH acid levels [6]. Some years ago, the catalytic mechanism of β-glucocerebrosidase was refined and allowed to identify the essential cofactor to the activity of this very enzyme. This is Saposine C (or SAP-C for Sphingolipid Activator Protein-C).

This way, some rare cases of Gaucher's disease were reported because of a defect in this coactivator, while the enzyme is synthetized normally by cells [7, 8].

I.4. Consequences of β-Glucocerebrosidase Activity Deficiency

Non-degraded glucosylceramid accumulate inside lysosomes of the reticuloendothelial system cells (macrophages) giving these cells a characteristic aspect with a crumpled cytoplasm with microscopy.

These cells called Gaucher's cells will infiltrate liver, spleen, and marrow bone inducing the main clinical signs of the disease: hepatomegalia, splenomegalia, osseous complications [9].

This way, according to the degree of cells' invasion, overload location (central nervous system implication or not) and the evolution speed of this tissular infiltration, various clinical forms of Gaucher's disease were individualized [6]:

- Gaucher Type 1, or chronic non-neurological (the most frequent) ;
- Gaucher Type 2, or acute neurological (the most serious but the less frequent) ;
- Gaucher Type 3, or subacute neurological (with intermediate seriousness in between the two other types).

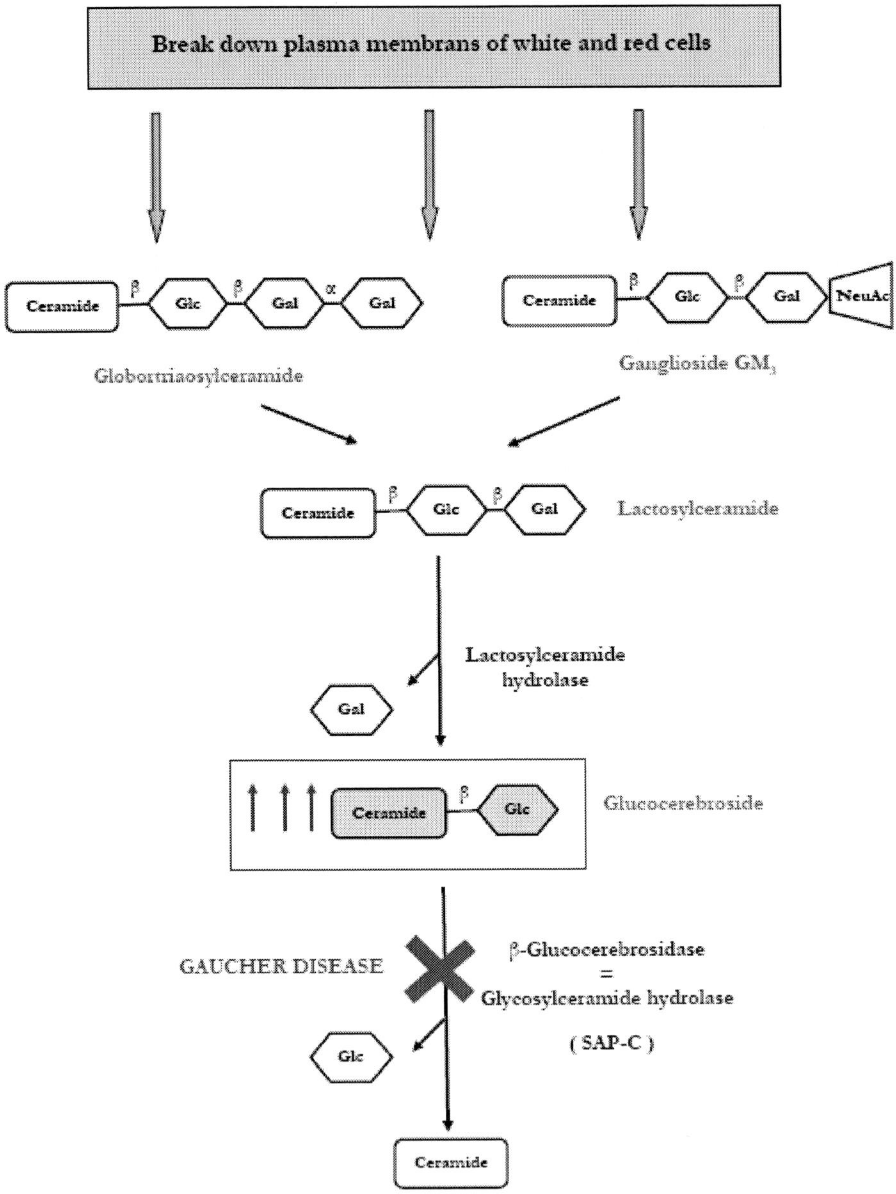

Figure 4. Site of deficient enzyme reaction in Gaucher's disease.

II. Our Experience in the Biochemical Diagnosis

II.1. Materials and Methods

In front of histological research lack of Gaucher's cell and the non-specificity of seric acid phosphatase, our laboratory has been keen on measuring the β-glucosidasic activity in leucocytes and splenic tissu in case of splenectomy.

II.1.1. Subjects Studied

Fifteen patients affected by the *non-neuropathic* clinical form of *Gaucher disease*, aged 4 to 13 years. The diagnosis was made clinically and confirmed in seven cases by the finding of typical Gaucher's cells in bone marrow.

Thirty five subjects, aged 4 to 15 years, without any clinical evidence or family history of Gaucher's disease, served as *controls.*

II.1.2. Isolation of Leukocytes and Spleen Cells

Leukocytes were isolated from venous blood collected with sodium heparinate and prepared as destribed by PETERS and al [10]. We have made some modifications as shown further in the discussion. If the patient have benefited by blood transfusion, it is important to perform the collection after 10 days.

Spleen, taken off from a patient with surgery (splenectomy), was immediately frozen and analysed within 48h as described by BRADY and al [11].

II.1.3. β-Glucocerebrosidase Activity Determination

To Characterize the biochemical defect in Gaucher disease, we study the cleaving potency of β-Glucocerebrosidase using synthetic fluoregenic substrate the 4-methylumbelliferyl-β-D-glucopyranoside (4Mu-β-glc) and sodium taurocholate as activator. The fluorescence of the enzymatically liberated 4-methylumbelliferone was measured, using excitation wavelenght of 365 nm and emission wavelenght of 450 nm [10].

Chromogenic substrates, the p-nitrophenyl-β-D-glucopyranoside and 2-hexadecanoylamino-4-nitrophenyl-β-D-glucopyranoside are widely available. However, these substrates are relatively non-specific, and a meticulous control of conditions is required for their accurate use.

We prefer to use the fluoregenic substrate 4Mu-β-glc which shows better sensitivity. As regards to the enzyme heterogeneity according to the physicochemical properties depending on the reactional environment, the measurement is carried out at 3,5 - 4 - 4,5 - 5 – 5,5 and 6 pH levels [12,13].m Protein level was determined with LOWRY and al method [14] using bovin serum albumin as standard.

II.1.4. Expression and Interpretation of the Results

- Activity = Nanomoles / mg of proteins /hour (leucocytes measurement) = Nanomoles / g of fresh weight /hour (splenic tissu measurment)

- Pourcentage of the deficiency always expressed by comparison to control groups launched at the same time with ills persons (pH curves).

II.2. Results

II.2.1. In the case of the splenictomized child, our results show (Figure 5):

- about the splenic tissu, whatever the reactional pH is, the ill person's activity collapses at optimum pH 5,5, it represents 24% of the control person activity.
- about the leucocytes of the same ill person the glucocerebrosidasic activity is constantly inferior to 30% at all pH levels. At optimum pH 5,5, it represents 17% of the control person activiy.

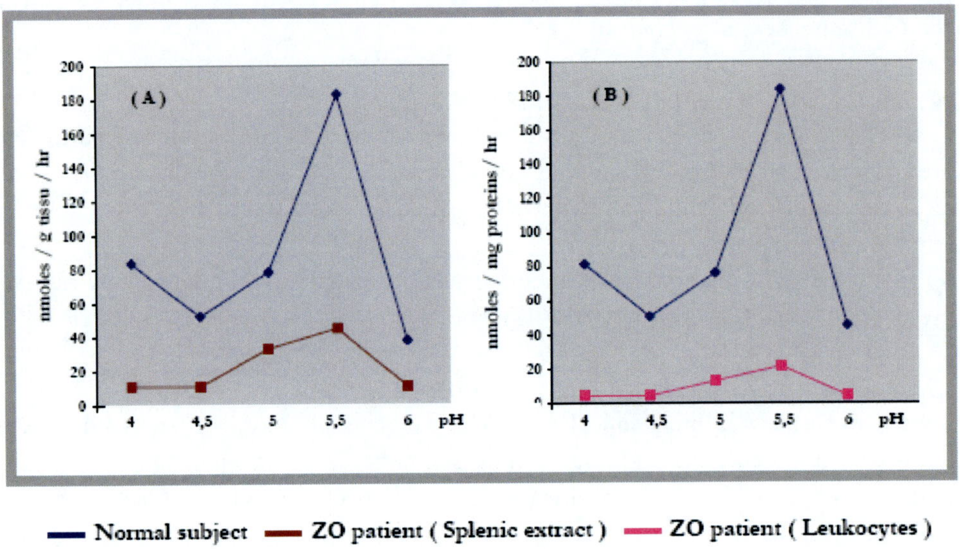

Figure 5. Representation pH-activity of β-glucocerebrosidase in splenic tissu (A) and leukocytes (B).

II.2.2. For the Rest of Ill Persons (13 Cases), The Enzymatic Deficiency Is Found at All pH Levels

It is particularly marked at pH 4 - 5 and 5,5 showing an activity ranging between 5 and 30 % compared to those of control persons (Figure 6).

II.2.3. Only One Ill Person Presented Subnormal Activities at Different pH Levels

The deficiency is found only at pH 4 with an activity equal to 15% in comparison to a control person (Figure 7).

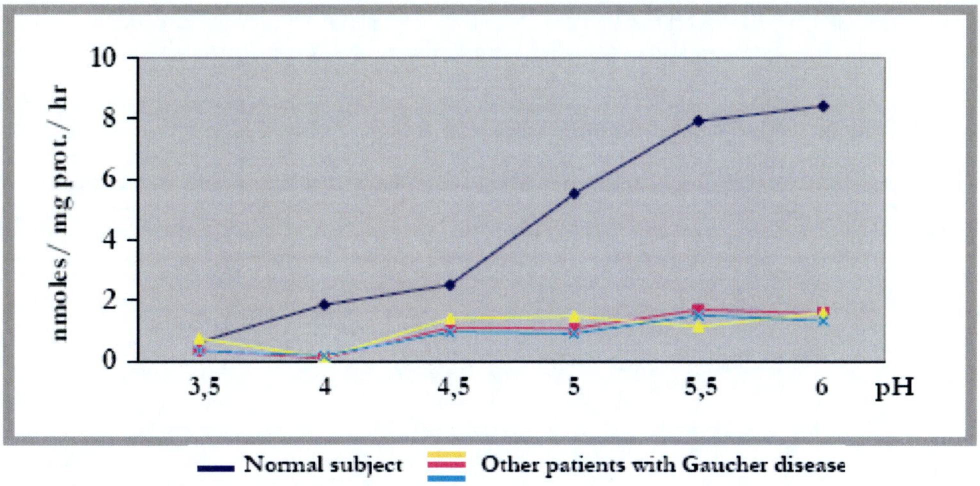

Figure 6. pH-activity curves for leukocyte β-glucocerebrosidase.

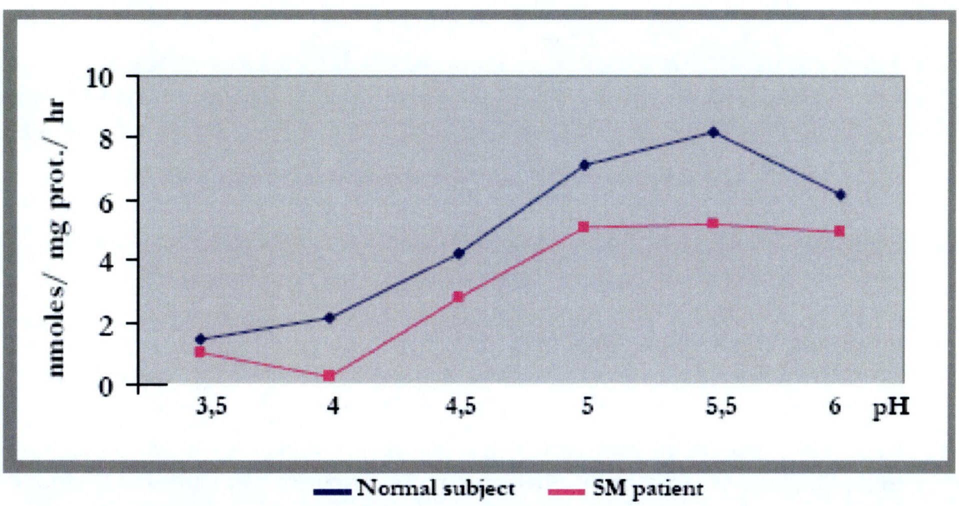

Figure 7. pH-activity curves for leukocyte β-glucocerebrosidase.

II.2.4. For Another Clinically Affected Person of Gaucher's Disease

The measurement has shown an activity of 38 % at pH 5, whereas the activity at other pH levels is superior to 50%. We recommended redoing the measurement. This has shown similar results but the lowest percentage stands at pH 4 (45 %) (Figure 8).

II.3. Discussion

II.3.1. β-Glucosidasic Activity Variability

Our results confirm the extreme heterogeneity of defects encountered in Gaucher's disease and therefore the necessity to perform this measurement at different reactional pH.

Indeed, after BEUTLER AND KUHL [15], Two intraleucocytary glucosidasic activities exist,one is maximum at pH 4, the other one at pH 5,5. What TURPIN and al bring aswell [16]. Besides, let's point out that optimal pH varies according to substrate type, buffers, detergents, aswell as, the lyse or not of cellular membranes.

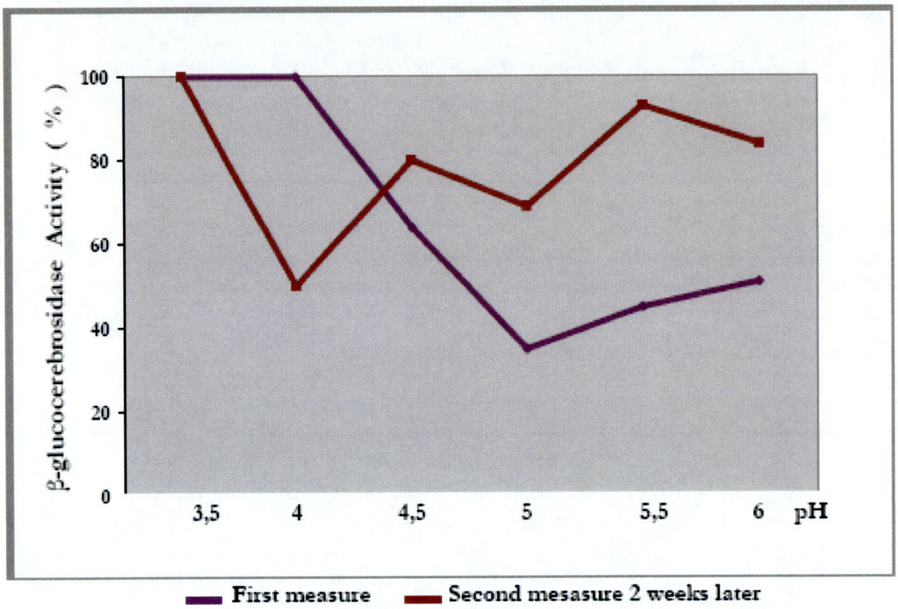

Figure 8. Representation pH-activity of β-glucocerebrosidase in leukocytes for YS patient.

Several modifications of the original technique of PETERS are currently used in order to assure a measurement at a unique optimal pH of 5,5.

They consist on:

- Lyphocytes separation (mononucleus cells) upon Ficoll, instead of total leukocytes. Indeed, the diagnosis is more reliable in mononucleus cells where acid β-glucosidase activity is more favourable;
- Pure taurocholic acid use instead of sodium impure taurocholate, which poses sometimes productivity problems;
- Cellular proteins adjustment (between 2 et 5 g/l) in leukocyts lysate, because a precise ratio between proteins quantity and detergent quantity required.

It is allowed that an intraleucocytar β-glucocerebrosidase activity inferior to 30 % assures the disease certainty of diagnosis. But what about patients who's activity is superior to this?

To remove all these unforeseens, the measurement must be repeated at least once.

Could a patient with a kept Gaucher's diagnosis have variable activities at different interval times?

Yes, intraleucocytar activity can vary from a pH to another for the same ill person on 02 different samples. This is due to the nature of leucocytar activity present in the extract to measure. Each cellular type can show an optimal activity at a particular pH depending on the isoenzyme present (Figure 9A).

Indeed, Peters and Lee's works have shown that several populations of leucocytes which profuseness in one or the other isoenzyme is variable. This suggests different preparations in order to obtain enriched extracts weather in monocytes, in lymphocytes, or in a mixed composition [17].

However, when the intraleucocytar activity is < 30 % at a particular pH, this very deficit is always found and confirmed at different interval times (Figure 9B).

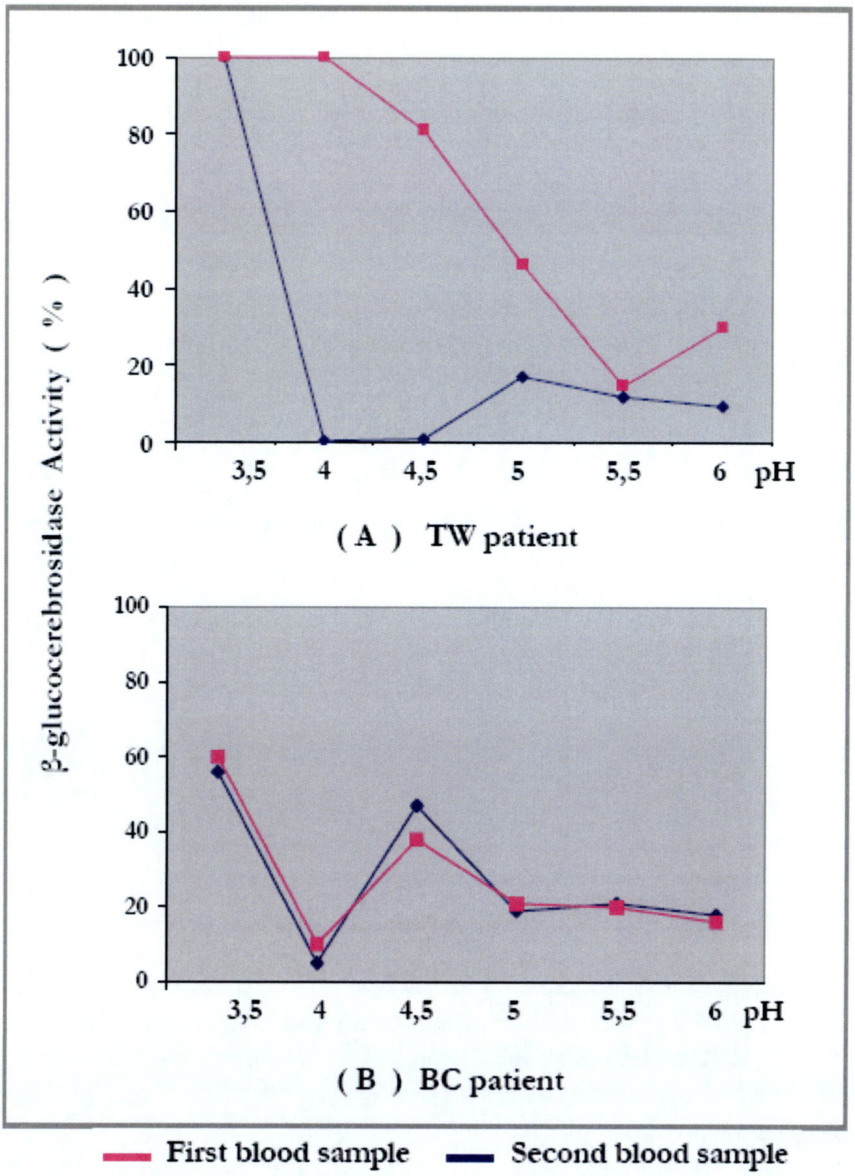

Figure 9. The chages in β-glucocerebrosidase's activity in leukocytes according to the pH and the blood sample collection.

II.3.2. Possibles Discordances between Measured Enzymatic Activity and Clinical Signs of Gaucher's Disease

Given, a caracteristic sysmptomatology of Gaucher's disease, three important elements must be pointed out:

a) the activity found in vitro may not reflect the activity in vivo. Indeed, the measurement realized use a synthetic substrate in physico chemical conditions dependant on the reactional environment.
b) the necessity to renew the measurement for a total confirmation. A second test is always recommended if the rate found, in the same laboratory to remove every doubt.
c) The clinical limits between Gaucher's disease and Niemann- Pick's disease (in the neurological form or not) are very non imprecise whereas the enzymatic deficiency is different. So, only the enzymatic determination of sphingomyelinase when the β-glucocerebrosidase activity is normal, can make the difference.

II.3.3. Treatment Supervision

Only one ill person in our series benefited by a substitutive treatment of *CEREZYME*. The enzymatic activity stayed unchanged (Figure 10).

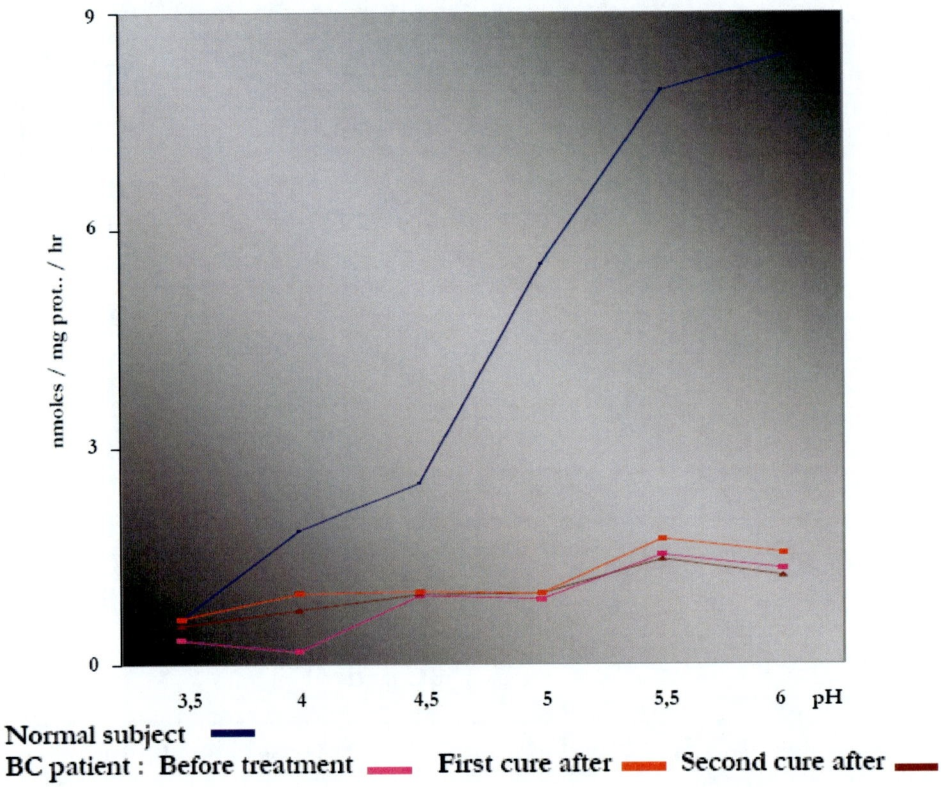

Figure 10. The enzymotherapy not effect the β-glucocerebrosidase's activity in leukocytes in Gaucher disease.

The is no point to measure the intraleucocytar activity of β-glucosidase but the the diagnosis confirmation. We never measure the β-glucosidase activity in order to supervise the therapy with CEREZYME.

For the supervision of the substitutive enzymotherapy in patients, we, currently recommend the measurement of 02 plasma enzymes:

- chitotriosidase;
- acid phosphatase tartrate-resistant (isoenzyme 5b).

Other markers such as angiotensine conversion enzyme (ACE), hexosaminidase and cytokines were recommended.

II.3.3.1. Plasma Chitotriosidase

Also, called: Chitinase; Chitodextrinase; 1,4-beta-poly-N-acetyl -glucosaminidase ; Poly-beta-glucosaminidase [18].

It is a good orientation marker because it is synthesized by activated macrophages of patients affected by Gaucher's disease.

Its activity is linked to glucosylceramide charge in the organism (in the reticuloendothelial system):

- activity ↑ 600 x N in 94 % of symptomatic patients of Gaucher's disease type 1 [19] ;
- normal activity to slightly enhanced :
- asymptomatic Gaucher's disease ;
- other lysosomal diseases ;
- comparable rise 10% of patients affected by β-thalassemia [20].
- nil activity in 6% of the population by the genetic deficiency of the enzyme.

It diminishes significantly after induction of *CEREZYME* by the cut of activated macrophages. Stopping treatment induces a rapid rise of this enzyme [21].

Its cutting represents an excellent marker of the treatment efficiency.

Measurement frequency recommended by the laboratory manufacturing *CEREZYME* is once every 3 months at the beginning of the treatment, than one every 12 months.

Chitotriosidase must find its place in certainty diagnosis of Gaucher's disease, as complement of β-glucosidase activity especially in the case where the deficiency is not inferior to 30%, in addition to its major interest in the treatment supervision. However, its measurement needs a fluorogenic substrate extremely costly that is not at our disposal.

II.3.3.2 Acid Phosphatase Tartrate Resistant (APTR) Type 5b

It is described as a marker enzyme of resorbant osteoclastes of bone.

It is known, aswell as phosphatase of type 5 and exists under the isoforms 5a et 5b. Only the isoform 5b of APTR is specifically associated with osteoclastes. It is produced by macrophages, too.

A monoclonal antibody recognizing the isoenzyme 5b of acid phosphates is used for the specific of measurement PATR [22].

The particularity of Gaucher's disease to rise significantly the APTRs (essentially the isoenzyme 5b) in serum and spleen. Certain studies set up a link between PATRs rise and the important osseous altering in the 03 clinical forms of Gaucher's.

After induction of *CEREZYME* , an important cutting of PATR 5b is associated to the osteolytic activity of Gaucher's cells in marrow bone and the improving of osseous remodelling [23].

II.3.4. Antenatal Diagnosis and Genetical Advice

An antinatal diagnosis of Gaucher's disease is possible because β-gluco- cerebrosidase is present in chorial villocities (10 - 12 weeks of amenorrhoea) and cultivated amniotic cells (16 weeks of amenorrhoea).

This diagnosis can be addressed to families who already got an index case.

II.3.5. Is the Molecular Diagnosis of Gaucher's Disease Justifiable?

Certainty diagnosis of Gaucher's disease can be made only by the measurement of β-glucocerebrosidase activity deficiency. The genotypic diagnosis allows to put in evidence the responsible mutations for the enzymatic deficiency and judge the disease seriousness according to the type found.

Glucocerebrosidase gene is located on the long arm of the chromosome 1.

More than 200 Mutations of the enzyme gene were identified, among them 5 more frequent : N370S - 84GC - L444P - IVS2 - Rec.

Mutations of the cofacteur SAP-c gene were recently identified [24].

Is the genetic therapy where the perspectives of the molecular diagnosis lie?

Conclusion

Given, the symptomatologic heterogeneity, the diagnosis of Gaucher's disease must appeal for a synergy between clear clinical signs and analytical means allowing certainty diagnosis.

Confirmation diagnosis can be assured only by the determination of β-glucocerebrosidase on leucocytes. This activity must be carried out at several pH levels ranging from 3,5 to 6 to identify the deficient isoenzymes, regarding the extreme heterogeneity of Gaucher's disease.

It is obvious that this measurement must be performed in laboratories with experience in the diagnosis of metabolic diseases.

Chitotriosidase is another parameter that appears not only a diagnostic element but a marker of therapeutic efficiency.

Patients affected by Gaucher's disease, currently benefit by a substitutive enzymotherapy. To measure the treatement efficiency, it is recommended to follow up seric enzymes action of lysosomal release, enzymes such as chitotriosidase.

Besides, the molecular diagnosis is possible by the research of mutations in question. These mutations are numerous and there only significance however stays far-off as far as an eventual genetic therapy is evoked.

Acknowledgements

I GRATEFULLY ACKNOWLEDGE THE ASSISTANCE AND ADVICE OF PR. A. BERHOUNE.

References

[1] GIESELMANN V. Lysosomal storage diseases *Biochim. Biophys. Actta* 1995 ; 1270 : 103-136.

[2] SANDHOFF K, KOLTER T. Biochemistry of glycosphingolipid degradation. *Clin. Chim. Acta* 1997 ; 266 : 51-56.

[3] BRADY RO, KANFER JN, SHAPIRO D. Metabolism of glucocerebrosides. II. Evidence of enzymatic deficiency in Gaucher's disease. *Biochem. Biophis. Res. Commun.* 1965 ; 18 : 221.

[4] PATRICK AD. A deficiency of glucocerebrosidase in Gaucher's disease. *Biochem. J.* 1965 ; 97 : 17c.

[5] HUTBERG B, OCKERMAN PA. β-Glucosidase activities in human tissues. Findings in Gaucher's disease. *Clin. Chim. acta* 1970 ; 283 : 169.

[6] BEUTLER E, GRABOWSKI GA. *Gaucher disease in the metabolic and molecular bases of inherited disease,* Ed. C. R. SCRIVER, A.L. BEAUDET, WS SLY. D VALLE, MC GRAW HILL, 1995 : 2641-2670.

[7] CHANG MH, BINDLOSS CA, GRABOWSKI GA, QI X, WINCHESTER B, HOPWOOD JJ. Saposins A, B, C, and D in plasma of patients with lysosomal storage disorders. *Clin. Chem.* 2000 ; 46 : 167-174.

[8] CONZELMAN E, SANDHOFF K. Activator proteins for lysosomal glycolipid hydrolysis. *Methods Enzymol.* 1987 ; 138 : 792-815.

[9] GERY I, ZIGLER JS, BRADY RO, BARRANGER JA. Selective effects of glucocerebroside (Gaucher's storage material) on macrophage cultures. *J. Clin. Invest* 1981 ; 68 : 1182-1189.

[10] PETERS SP, LEE RE, GLEW RH. A micro assay for gaucher's disease. *Clin. Chim. Acta* 1975 ; 60 : 391-396.

[11] BRADY RO, KANFER JN, SHAPIRO D. The metabolism of glucocerebrosides. I. Preparation and properties of glucocerebroside cleaving enzyme from spleen tissue. *J. Biol. Chem.* 1966 ; 240 : 39-43.

[12] YARGUI L, BERHOUNE A. A propos de la maladie de Gaucher retrouvée chez un adulte : Intérêt de la β-glucosidase leucocytaire. In : VII° Journées Nationales d'enseignement post-universitaire en Pharmacie. *Algiers*, 24-25 mai. 2000.

[13] YARGUI L, MOKHTARI S, ARAB M, BERHOUNE A. The biochemical diagnosis of Gaucher disease. *Arch. Ped.* 2006 ; 12: 1301-1309 (article in french).

[14] LOWRY OH, ROSENBOROUGH NY, FARR AL, RANDALL RJ. Protein measurements with the folin-phenol reagent. *J. Bio. Chem.* 1951 ; 193 : 265-275.

[15] BEUTLER E, KUHL W. The diagnostic of the adult type of Gaucher's disease an dits carrier state by demonstration of deficiency of β-glucosidase activity in peripheral blood leukocytes. *J. Lab. Clin. Med.* 1970 ; (76) 5 : 747-755.

[16] TURPIN JC, BAUMAN N, CASTAIGNE P. Diagnostic biologique de la maladie de Gaucher. *Nouv. Press Med.* 1972 ; 1 : 1579-1581.

[17] KARAZEH A, CARROLL M. Physical and Kinetic Properties of β-glucosidase in Gaucher disease. *J. Inher. Metab. Dis.* 1983 ; 6 : 101-104.

[18] ARONSON NN, HALLORAN BA, ALEXYEV MF, AMABLE L, MADURA JD, PASUPULATI L, WORTH C, VAN ROEY P. Family 18 chitinase–oligosaccharide substrate interaction: subsite preference and anomer selectivity of Serratia marcescens chitinase A. *Biochem. J.* 2003 ; 376 : 87–95.

[19] HOLLAK CE, VAN WEELY S, VAN OERS MH, AERTS JM. Marked elevation of plasma chitotriosidase activity. A novel hallmark of Gaucher Disease. *J. Clin. Invest.* 1994 ; 93 : 1288-1292.

[20] BARONE R, DI GREGORIO F, ROMEO MA, SCHILIRO G, PAVONE L. Plasma Chitotriosidase Activity in Patients with β-thalassemia. *Blood Cells, Molecules and Diseases* 1998 ; 25 :1-8.

[21] GUO Y, WANG H, BOER AM, WEVERS RA, DE BRUIJN AM, GROENER JEM, HOLLAK CEM, AERTS JMFG, GALJAARD H, VAN DIGGELEN OP. Elevated plasma chitotriosidase activity in various lysosomal storage disorders. *J. Inher. Metab. Dis.* 1995; 18 : 717-722.

[22] HANNON RA, EAGLETON AC, CLOWES JA, MACHADO ABC, REASTELL R, BLUMSOHN A. Clinicalevaluation of a novel activity-based immunoessay for tartrate resistant acid phosphatase isoform 5b. In : International Osteoporosis Foundation. World Congress on Osteoporosis. *Lisbonne*, 11-15 mai. 2002.

[23] TROY K, CUTTNER J, REILLY M, GRABOWSKI G, DESNICK R. Tartrate resistant acid phosphatase staining of monocytes in Gaucher disease. *Am. J. Hematol.* 1985; 19: 237-244.

[24] CAILLAUD C, POENARU L. Maladies de Gaucher et de Fabry: aspects biochimiques et génétiques. *J. Soc. Bio.* 2002; 196 (1) : 135-140.

In: Glycolipids: New Research
Editor: Daiki Sasaki, pp. 19-28

ISBN: 978-1-60456-216-3

Short Communication B

Modulation of Fatty Acid Molecular Species in Sulfatide and Its Precursor, Galactosylceramide in Madin-Darby Canine Kidney Cells under Osmotic Stresses[#]

Yukio Niimura[1*] and Ken-ichi Nagai[2]

[1]Research Center of Biomedical Analysis and Radioisotope,
Faculty of Medicine
[2]Department of Laboratory Medicine, Faculty of Medical Technology,
Teikyo University, 2-11-1 Kaga Itabashi-ku,
Tokyo, Japan, 173-8605

Abstract

It is known that the concentration of sulfatide (SM4s), a major sulfoglycolipid in Madin-Darby canine kidney (MDCK) cells increases in a hyperosmotic medium due to the activation of the cerebroside sulfotransferase gene. We investigated the changes in molecular species of sulfoglycolipids in this cell line under osmotic stresses. Analysis by thin layer chromatography showed that the slower-moving band of SM4s specifically increased in hyperosmotic medium. Moreover, analysis by negative ion liquid secondary ion mass spectrometry demonstrated an increase in the ratios of hydroxy fatty acid species to those of corresponding non-hydroxy fatty acid ones of SM4s, under hyperosmotic stress. Further, analysis of galactosylceramide, the precursor of SM4s, by negative ion LSIMS also showed a similar result, which suggests that the hydroxylation of fatty acids in sulfatide was specifically induced through the hydroxylation of its

[#] The nomenclature system for lipids follows the recommendation of the Nomenclature Committee, International Union of Pure and Applied Chemistry (IUPAC-IUB, 1999) [1].

[*] To whom correspondence should be addressed: Tel.: +81-3-3964-3479, Fax: +81-3-3964-5807; E-mail: yniimura@med.teikyou.ac.jp

precursor glycolipid, under hyperosmotic stress. The elucidation of the mechanism that increases the hydrophilic molecular species in galactosphingolipids of renal cells under hyperosmotic stress may contribute to the *in vitro* study on the function of galactolipids containing hydroxy fatty acids.

Abbreviations

Cer	ceramide;
d18:1,	4-sphingenine;
LSIMS	liquid secondary ion mass spectrometry;
GalCer	galactosylceramide;
GlcCer	glucosylceramide;
SM4s	sulfatide.

Introduction

Sulfatide (SM4s) is a major component of myelin in the vertebrate brain, whereas there are various types of sulfoglycolipids including sulfatides ubiquitously distributed in the vertebrate kidney [2]. Recently, we characterized many types of sulfoglycolipids from African green monkey kidney cell line [3]. It has been shown that Madin-Darby canine kidney (MDCK) cells synthesize sulfated glycosphingolipids, SM4s (GalCer I^3-sulfate) and SM3 (LacCer II^3-sulfate), and their synthesis depends on the osmolality of the medium [4, 5]. Further, the hyperosmosis-resistant clones selected from MDCK cells accumulated these sulfoglycolipids remarkably [6]. SM4s appears to have accumulated more abundantly in the medulla than in the cortex of the kidney [7]. Mammalian kidney cells are exposed to various osmotic conditions during the process of urine concentration and are well adapted to both hypotonic and hyperosmotic stresses. The allometric relationship between the sulfoglycolipid concentration in the kidney and the body weight suggests that sulfoglycolipids are important for water transport in the kidneys [8]. These facts suggest that sulfoglycolipids in the renal epithelial cell membrane act as a membrane barrier against various osmotic conditions during the process of osmoregulation. Recently, we also reported a selective increase in the levels of some molecular species of SM4s in MDCK cells, under hyperosmotic stress. The specific response of glucosylceramide sulfate, a unique sulfoglycolipid, to hypotonic stress was also revealed [9]. In this study, we report the specific changes that occur in the fatty acid molecular species of sulfoglycolipid (SM4s) and its precursor glycolipid (galactosylceramide), under hyperosmotic stress.

Materials and Methods

Cell culture MDCK/P3 cells were obtained from the JCRB Cell Bank (Tokyo, Japan) and were cultured in Dulbecco's modified Eagle's medium containing 10% fetal bovine serum.

For osmotic stress experiments, cells were cultured in PM1000 medium (300 mOsm/l, Eikenkagaku, Tokyo, Japan) containing insulin (10 μg/ml), without serum. The previously described osmotic stresses were applied [5].

Preparation of sulfoglycolipids and neutral glycolipids Sulfoglycolipids were separated by chloroform/methanol extraction, mild alkali treatment, DEAE-Sephadex A-25 column chromatography, and thin layer chromatography (TLC), as described previously [3]. TLC was performed on Silica Gel 60 high-performance TLC (HPTLC) plates (E. Merck, Darmstadt, Germany). Sulfoglycolipids were separated on TLC using chloroform/methanol/acetone/acetic acid/water (8:2:4:2:1, v/v) as the solvent. Neutral glycolipids were developed on TLC twice, first using chloroform/methanol/water (60:25:4, v/v) and second using propanol/ammonium hydroxide/methyl acetate/water (15:2:1:3, v/v), and monohexosyl ceramide fractions were obtained from the HPTLC plate. The glycolipid bands were visualized using iodine vapors or by spraying the plate with orcinol-H_2SO_4 reagent and heating it for 5 min at 120°C. SM4s, SM3, and GlcCer-I^3-sulfate were obtained from the rat kidney in our laboratory and used as standards.

Mass spectrometry Negative ion liquid secondary ion mass spectrometry (LSIMS) was performed using the Concept 1H spectrometer (Shimadzu/Kratos, Kyoto, Japan) fitted with a cesium ion gun [10]. Approximately, 0.5 nmol of underivatized glycolipids were mixed with 1 μl triethanolamine, which was used as the matrix. The spectra were recorded at an accelerating voltage of 8 kV with a scan rate of 5 s/decade and at a resolution varying from 1000–2000.

Results

Analysis of Sulfatide Molecular Species by Negative Ion LSIMS

SM4s of MDCK cells cultured in normal medium separated into two main bands on TLC, when subjected to the solvent containing chloroform/methanol/acetone/acetic acid/water (8:2:4:2:1, v/v). The upper fast-moving band and the lower slow-moving band isolated from TLC plate were analyzed by negative ion LSIMS (Figure 1). The upper band showed the major fatty acid molecular species of m/z 890 (24:0), 888 (24:1), 862 (22:0), and 860 (22:1) in combination with d18:1 as a sphingosine (Figure 1A). The lower band showed three major peaks of m/z 906 (24h:0), 878 (22h:0), and 778 (16:0) (Figure 1B).

The non-hydroxy fatty acids species were detected mainly in the upper band on TLC plate by negative ion LSIMS, whereas the hydroxy fatty acids and the shorter chain length species were detected in the lower band. Whole fraction of SM4s, obtained from MDCK cells cultured in both the control and hyperosmotic media was analyzed by negative ion LSIMS (Figure 2A and B). The molecular species of non-hydroxy fatty acids (C16:0, C16:1, C18:0, C20:0, C20:1, C22:0, C22:1, C24:0, and C24:1) and their corresponding hydroxy fatty acids in combination with d18:1 as a sphingosine were detected in both the samples. Ratio of the hydroxy fatty acid species to the non-hydroxy fatty acid ones was observed (Figure 2) and appeared to increase in the longer chain fatty acids (>C18:0) of SM4s under hyperosmotic stress.

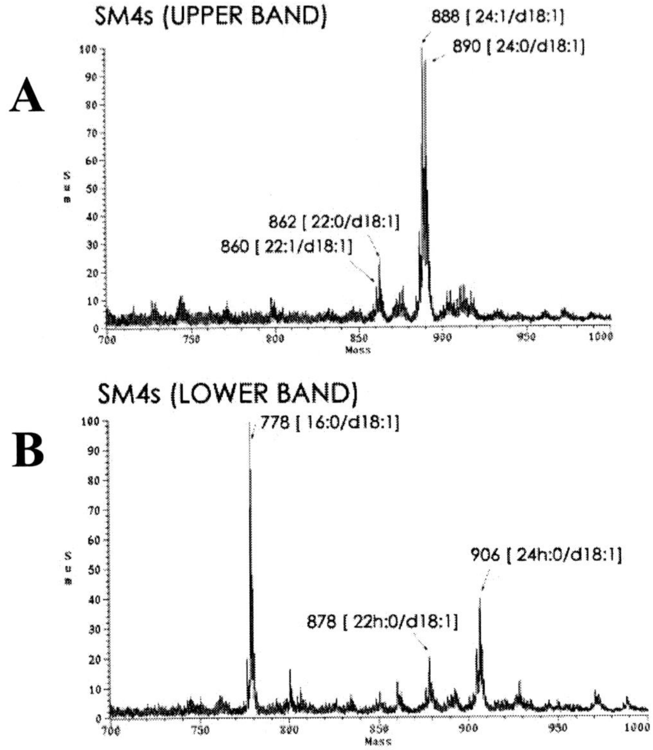

Figure 1. Mass spectra in partial fraction of sulfatide. Cells were cultured in the control medium for 2 days. Partial fraction of sulfatide was obtained from the preparative TLC. A: Upper band fraction of SM4 from control culture, B: Lower band fraction of SM4s from control culture.

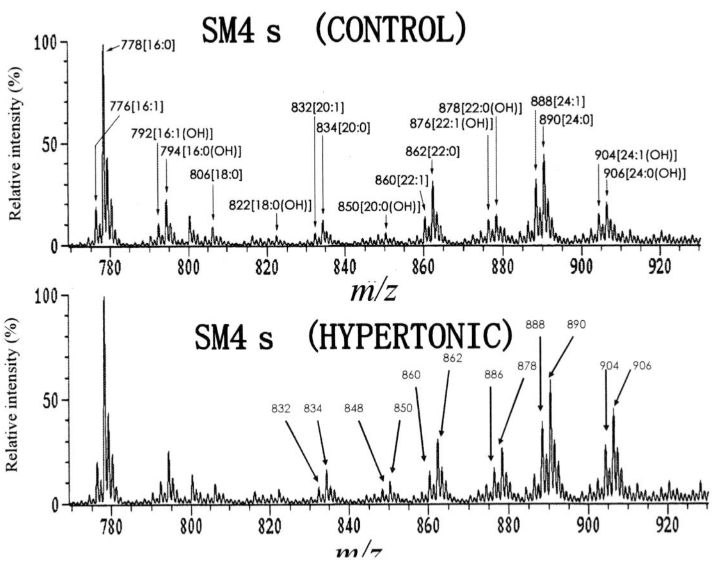

Figure 2. Mass spectra in whole fraction of sulfatide. Cells were cultured either in the control or hyperosmotic medium (500 mOsm/l), which was prepared by adding NaCl for 2 days. Whole fraction of sulfatide was obtained from the preparative TLC. A. SM4s from control culture, B: SM4s from hyperosmotic culture.

From the above data, we found that the hydroxy fatty acid species of SM4s increased when the culture was grown in hyperosmotic medium, with a corresponding increase of SM4s in the lower band on TLC. The ratio of hydroxy fatty acids to non-hydroxy fatty acids, with the exception of short fatty acid (C16:0) increased in the hyperosmotic culture (Figure 3).

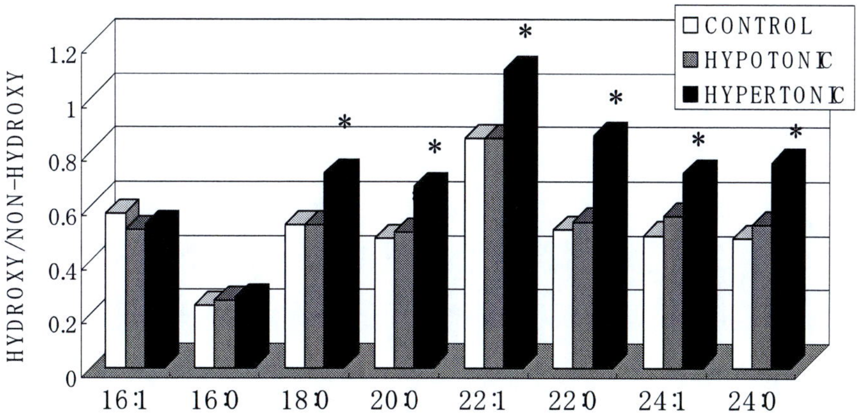

Figure 3. Ratio of the hydroxy fatty acids-containing species of SM4s to the non-hydroxy fatty acid ones. Ratios were obtained from the peak heights in both hydroxyl and non-hydroxy fatty acid molecular ions, determined by negative ion LSIMS as shown in Figure 2. Significance level (*) to the control was below 5%.

SM3, shown as a single band on TLC, was also isolated and analyzed by LSIMS. In the control culture, fatty acid molecular species of m/z 1068 (24h:0), 1052 (22:0), 1040 (22h:0), 1024 (22:0), 996 (20:0), 956 (16h:0), and 940 (16:0) in combination with d18:1 as a sphingosine were observed (Figure 4A).

The hydroxy fatty acid molecular species of SM3 from the cells were unaltered by the culture conditions in both the control and hyperosmotic media (Figure 4B).

Glucosylceramide sulfate, detected as a minor sulfoglycolipid in MDCK cell, was also analyzed in fatty acid molecular species. The results did not show any difference in hydroxy fatty acid species under hyperosmotic stress (data not shown).

Effect of osmotic stresses on the molecular species of monohexosyl ceramide We also analyzed the precursor glycolipids of these sulfoglycolipids. The neutral glycolipid fraction was separated from the acidic fraction by eluting through the DEAE-Sephadex column. The neutral glycolipids that were purified using two-dimensional TLC were then compared. In the control culture, glucosylceramide (GlcCer, 2 spots) and galactosylceramide (GalCer, 2 spots) were separated. The slow-moving spot of GalCer showed remarkable accumulation in the hyperosmotic culture, while both spots of GalCer showed reduced accumulation levels in the hypotonic culture. The faster-moving spot of GlcCer showed increased levels in the hypotonic culture. Then, GlcCer and GalCer were separated on TLC plates and their molecular species were identified by LSIMS in a manner similar to SM4s.

The ratios of the hydroxy fatty acid species to the non-hydroxy fatty acid ones in GlcCer and GalCer were also determined in a manner similar to SM4s (Figure 5).

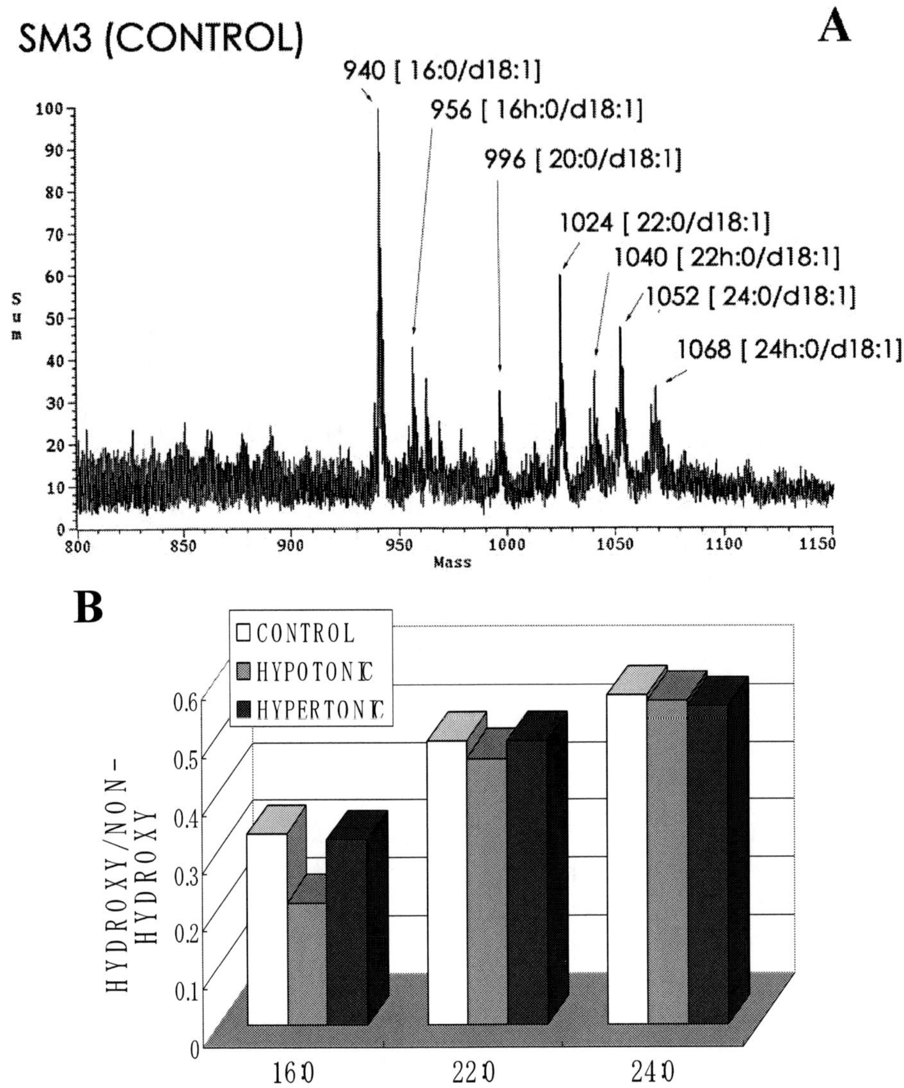

Figure 4. Mass spectra of SM3 and the changes of fatty acid molecular species by osmotic stresses. SM3 was prepared from the HPTLC plate as described under "Materials and Methods." A: SM3 from control culture, B: Ratios were obtained from the peak heights in both hydroxyl and non-hydroxy fatty acid molecular ions of SM3, determined by negative ion LSIMS.

The ratios of the hydroxy fatty acid species to the non-hydroxy fatty acid ones in GalCer increased, while those in GlcCer remained unchanged, under hyperosmotic culture conditions. A slight decrease in the ratio of these fatty acids in GlcCer was observed under hypotonic conditions. Enhancement of the polar GalCer in the hyperosmotic culture corresponded to the increase of the slow-moving spot of GalCer on TLC (data not shown).

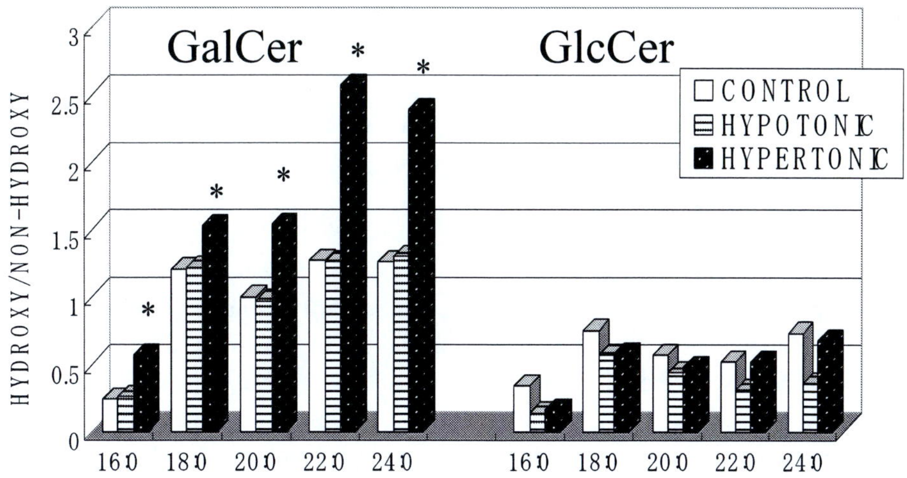

Figure 5. Changes of molecular species in monohexosyl ceramides. GalCer and GlcCer were prepared from the HPTLC plate as described under "Materials and Methods". Ratios were obtained from the peak heights in both hydroxyl and non-hydroxy fatty acid molecular ions of GalCer and GlcCer, determined by negative ion LSIMS. Significance level (*) to the control was below 5%.

Discussion

The accumulation of sulfoglycolipids SM4s and SM3 in MDCK cells under hyperosmotic stress was due to the elevation of the cerebroside sulfotransferase activity [11], and the reduction in its degradation rate [6]. In a recent report, it was shown that the increase in the level of SM4s in MDCK cells under hyperosmotic stress was induced by both, an increase in the precursor (GalCer) and the activation of the cerebroside sulfotransferase gene [9]. During this study, we also found changes in the intensity of the slow-moving band of SM4s fraction on TLC, under hyperosmotic conditions [5]. The slow-moving band on TLC, suggesting the presence of more hydrophilic species in SM4s or galactosylceramide, stained more intensely under hyperosmotic stress than under the control conditions. We reported the differences in fatty acid composition of these galactosphingolipids by negative ion LSIMS and confirmed the increase of more hydrophilic hydroxy fatty acid species under hyperosmotic stress conditions. In this study, we found that the molecular species of SM4s was modified under the hyperosmotic conditions leading to an increase in the levels of hydroxy fatty acids. We also found that the hydroxy fatty acid species accumulated in the precursor galactolipids (GalCer) under the same hyperosmotic stress, but not in SM3 and glucosylceramide sulfate. This is the first report regarding the *in vitro* changes in fatty acid molecular species of galactosphingolipids in renal cells. Galactosphingolipids such as GalCer and SM4s are ubiquitously distributed in the vertebrate brain and kidney. The galactosphingolipids in these tissues are predominantly composed of the hydroxy fatty acid molecular species, while the other glycolipids are primarily composed of non-hydroxy fatty acids. In the brains of mammals, 2-hydroxy sphingolipids are uniquely abundant in GalCer and SM4s. A human gene of a fatty acid 2-hydroxylase (*FA2H*) has also been characterized

[12] and is observed to be highly expressed in the brain. Fatty acid 2-hydroxylase forms free 2-hydroxy fatty acids, which act as precursors of myelin 2-hydroxy galactolipids. The fatty acid composition of galactolipids in neonatal mouse brain has been observed to gradually increase during the course of myelination [13]. The time course of its expression in the brain closely follows the expression of myelin-specific genes, reaching a maximum at 2-3 weeks of age [14].

During the process of keratinocyte differentiation, the fatty acid 2-hydroxylase, encoded by *FA2H* gene, accounts for an increase in the levels of 2-hydroxy ceramides [15]. 2-Hydroxy ceramides and 2-hydroxy glucosylceramides are required for epidermal lamellar membrane formation. Thus, the differentiation-linked increases in *FA2H* expression are essential for epidermal permeability barrier homeostasis. In renal cells, we assumed that the hyperosmotic stress would probably activate *FA2H*. This would selectively increase the production of hydroxy fatty acids, and accelerate the synthesis of hydroxy fatty acids ceramide and galactosylceramide by galactosyltransferase, and finally hydroxy fatty acid SM4s by cerebroside sulfotransferase.

The deficiencies of GalCer and SM4s in the mouse brain lead to an increased fluid permeability and impaired packing of the myelin lipid bilayer in the internodal membrane system. The loss of the two glycosphingolipid classes would cause a breakdown of saltatory conductance of myelinated axons in the knockout mouse [16]. Although, little is known about the significance of hydroxy fatty acid molecular species of galactolipids in kidneys, it is a possibility that they enhance the hydroxy bonds in the bilayer of the plasma membrane and contribute to the modification of membrane fluidity. This would possibly improve the functioning of renal cell membrane as a membrane barrier in the epidermis. These changes may also affect the properties of the cell membrane, which adapts to the osmotic stress. Both GalCer and SM4s are known to accumulate at the basolateral surface of MDCK cells, while GlcCer is preferentially transported to the apical cell surface [17, 18]. The membrane polarity of the renal epithelial cell is considered to play a crucial role for the transportation of various substances in the kidney. Hyperosmotic stress may promote the differentiation of basolateral cell surface of renal epithelia. This cell line may serve as an excellent model for studying the membrane polarity in renal epithelial cells. Recently, we reported the existence of glucosylceramide sulfate as a non-galactolipid in monkey kidney cell lines (Verots S3) [3] and MDCK/P3 cells [9]. Glucosylceramide sulfate was first isolated and characterized from rat kidney [19]. We found that the glucosylceramide sulfate, a non-galactolipid, accumulated without significant difference in the hydroxy fatty acid species under hypotonic conditions. The galactosylceramide sulfotransferase gene has been cloned from renal cancer cells [20]. We think the cerebroside sulfotransferase in kidney may be responsible for the sulfation of many types of sulfoglycolipids [3]. Considering the fact that these sulfoglycolipids would be synthesized solely by the sulfotransferase, the inverse relationship between the increase in the levels of glucosylceramide sulfate and the reduction in the levels of other sulfoglycolipids under hypotonic stress would be due to the increased supply of their precursors. In kidney of galactosylceramide synthase-deficient mice, there was an increase in the levels of more polar sulfoglycolipids in compensation to the absence of sulfatide [21]. The results of the study using cerebroside sulfotransferase-deficient mice, showed SM4s to be a major L-selectin-binding molecule in the kidney [22]. Although these sulfoglycolipids are not the major components required for normal kidney cellular operations, they still play crucial roles in raising quality of renal function. Although, the hyperosmotic stress increased ceramide level in renal cells, apoptosis was not induced. Thus, the role of ceramide as a mediator in kidney still remains to be elucidated.

References

[1] IUPAC-IUB, (1999) Joint Commission on biochemical nomenclature (JCBN). Nomenclature of glycolipids: Recommendations 1997. *Glycoconj. J.*, *16*, 1-6.

[2] Ishizuka, I. (1997) Chemistry and functional distribution of sulfoglycolipids. *Prog. Lipid Res. 36,* 245-319.

[3] Niimura, Y. and Ishizuka, I. (2006) Isolation and identification of nine sulfated glycosphingolipids containing two unique sulfated gangliosides from the African green monkey kidney cells, Verots S3, and their possible metabolic pathways. *Glycobiology*, *16*, 729-735.

[4] Niimura, Y. and Ishizuka, I. (1986) Glycosphingolipid composition of a renal cell line (MDCK) and its ouabain-resistant mutant. *J. Biochem. 100*, 825-835.

[5] Niimura, Y. and Ishizuka, I. (1990) Adaptive changes in sulfoglycolipids of kidney cell lines by culture in anisosmotic media. *Biochim. Biophys. Acta 1052*, 248-254.

[6] Niimura, Y. and Ishizuka, I. (1991) Accumulation of sulfoglycolipids in hyperosmosis-resistant clones derived from the renal epithelial cell line MDCK (Madin-Darby canine kidney cell). *Comp. Biochem. Physiol. 100B*, 535-541.

[7] Karlsson, K.A., Samuelsson, B.E. and Steen, G.O. (1973) The sphingolipid composition of bovine kidney cortex, medulla and papilla. *Biochim. Biophys. Acta 316*, 317-335.

[8] Nagai, K.I., Ishizuka, I. and Oda, S.I. (1984) Acidic glycolipids from kidney of suncus (*Insectivora*). *J. Biochem. 95*, 1501-1511.

[9] Niimura Y. and Nagai, K.I. (2007) Metabolic responses of sulfatide and related glycolipids in Madin-Darby canine kidney (MDCK) cells under osmotic stresses. *Comp. Biochem . Physiol. B* in press.

[10] Tadano-Aritomi, K., Kubo, H., Ireland, P., Hikita, T. and Ishizuka, I. (1998) Isolation and characterization of a unique sulfated ganglioside, sulfated GM1a, from rat kidney. *Glycobiology, 8*, 341-350.

[11] Okuda, M. (1995) The sulfoglycolipid composition and activity of glycolipid sulfotransferase in the cultured renal cell line (Madin-Darby canine kidney cell) are altered by NaCl concentration. *Teikyo Medical J. 18*. 209-219.

[12] Alderson, N.L., Rembiesa, B.M., Walla, M.D., Bielawska, A., Bielawski, J. and Hama, H. (2004) The human FA2H gene encodes a fatty acid 2-hydroxylase. *J. Biol. Chem. 279*, 48562-48568.

[13] Alderson, N.L., Maldonado, E.N., Kern, M.J., Bhat, N.R. and Hama, H. (2006) FA2H-dependent fatty acid 2-hydroxylation in postnatal mouse brain. *J. Lipid Res. 47*. 2772-2780.

[14] Eckhardt, M., Yaghootfam, A., Fewou, S.N., Zöller, I., and Gieselmann, V. (2005) A mammalian fatty acid hydroxylase responsible for the formation of alpha-hydroxylated galactosylceramide in myelin. *Biochem J. 388,* 245-54.

[15] Uchida, Y., Hama, H., Alderson, N.L., Douangpanya, S., Wang, Y., Crumrine, D.A., Elias, P.M., and Holleran, W.M. (2007). Fatty acid 2-hydroxylase, encoded by FA2H, accounts for differentiation-associated increase in 2-OH ceramides during keratinocyte differentiation. *J Biol Chem. 282*, 13211-13219.

[16] Bosio, A., Binczek, E., Haupt, W.F. and Stoffel, W. (1998) Composition and biophysical properties of myelin lipid define the neurological defects in galactocerebroside- and sulfatide-deficient mice. *J Neurochem. 70,* 308-15.

[17] van der Bijl, P., Lopes-Cardozo, M. and van Meer, G. (1996) Sorting of newly synthesized galactosphingolipids to the two surface domains of epithelial cells. *J. Cell Biol. 132*, 813-821.

[18] Nichols, G.E, Shiraishi, T. and Young, W.W.Jr. (1988) Polarity of neutral glycolipids, gangliosides, and sulfated lipids in MDCK epithelial cells. *J. Lipid Res. 29*, 1205-1213.

[19] Iida, N., Toida, T., Kushi, Y., Handa, S., Fredman, P., Svennerholm, L. and Ishizuka, I. (1989) A sulfated glucosylceramide from rat kidney. *J. Biol. Chem. 264*, 5974-5980.

[20] Honke, K., Tsuda, M., Hirahara, Y., Ishii, A., Makita, A. and Wada, Y. (1997) Molecular cloning and expression of cDNA encoding human 3′-phosphoadenylylsulfate:galactosylceramide 3′-sulfotransferase. *J. Biol. Chem., 272*, 4864-4868.

[21] Tadano-Aritomi, K., Hikita, T., Fujimoto, H., Suzuki, K., Motegi, K. and Ishizuka, I. (2000) Kidney lipids in galactosylceramide synthase-deficient mice. Absence of galactosylsulfatide and compensatory increase in more polar sulfoglycolipids. *J. Lipid Res. 41*, 1237-1243.

[22] Ogawa, D., Shikata, K., Honke, K., Sato, S., Matsuda, M., Nagase, R., Tone, A., Okada, S., Usui, H., Wada, J., Miyasaka, M., Kawashima, H., Suzuki, Y., Suzuki, T., Taniguchi, N., Hirahara, Y, Tadano-Aritomi, K., Ishizuka, I., Tedder, T.F. and Makino, H. (2004) Cerebroside sulfotransferase deficiency ameliorates L-selectin-dependent monocyte infiltration in the kidney after ureteral obstruction. *J. Biol. Chem. 279*, 2085-2090.

In: Glycolipids: New Research
Editor: Daiki Sasaki, pp. 29-52
ISBN: 978-1-60456-216-3

Chapter I

Gangliosides' Role in Cellular Interactions with Microbes and Immune Cells Activation

Luz P. Blanco*
Department of Internal Medicine
Division of Allergy and Clinical Immunology
Michigan Nanotechnology Institute for Medicine and Biological Sciences
9346 MSRB III, 1150 W Medical Ctr. Dr.
University of Michigan, Ann Arbor MI 48109-0648
Reviewed by: Christiane Wobus
Department of Microbiology and Immunology,
Medical School, University of Michigan.

Abstract

Glycolipids—in particular gangliosides—are ubiquitous molecules on the surface of cells. They play roles in important cellular processes like differentiation and transformation. This chapter focuses on the role of gangliosides in the interaction of eukaryotic cells with microbes and microbe-derived elements, such as toxins, as well as the role of gangliosides in the activation of immune cells. Gangliosides can be considered cellular gateways for certain microbes and viruses that have the ability to interact and to invade eukaryotic cells. Most of these early interactions during infection initially take place in the lipid raft domains of the plasma membrane. Gangliosides also regulate the immune system and play a role in the efficacy of certain experimental mucosal adjuvants (i.e., substances that enhance the immune response against antigens). However, their mechanism of action is not yet well elucidated. Consequently, the outcome of some host-pathogen interactions may be manipulated at two distinct levels: at the initial host-pathogen interaction and through intervention at the specific immune response that is induced later.

* Phone number: 734-647-0052; Fax number: 734-936-2990; lblanco@umich.edu

Moreover, gangliosides serve as receptors for some host molecules, such as antibodies. In addition, their attachment can lead to the recruitment and activation of toll-like receptors and signaling transduction pathways that promote interactions and activate diverse type of cells. This chapter will explore the participation of gangliosides and lipid rafts in these diverse events. A better understanding of the molecular mechanism involved in these interactions will allow us to manipulate host-pathogen and host immune responses more effectively to prevent infectious diseases and to optimize vaccination and prophylaxis against certain diseases.

The Participation of Gangliosides and Lipid Rafts in Host-Pathogen Interactions

Lipid rafts are membrane domains which are enriched in cholesterol and glycosphingolipids. Diverse cellular activation processes leading to cellular transformation and differentiation are initiated in the lipid domains [68]. Furthermore, many diverse microbes such as bacteria, parasites, and viruses initiated infections of the eukaryotic cells through interaction with the lipid raft domains [109, 66]. Lipid rafts per se—as well as specific components of lipid rafts such as gangliosides, ceramides and cholesterol—play a role in the adherence and internalization by eukaryotic cells of several viruses and bacteria as depicted in Table 1; even some parasites use lipid rafts to invade cells. What is so special about lipid rafts' environment is that receptors coalesce when clusters of lipid rafts are formed. The clustering in these platforms then triggers and facilitates signal transduction processes that require the physical proximity of functional molecules, resulting ultimately in membrane and cellular modifications (such as ruffling and cellular activation). In fact, the glycosynapse concept developed by Hakomori supports the proposition that gangliosides play a functional role in host-pathogens interactions [40]. Hakomori proposed that gangliosides in clusters or microdomains, together with functional membrane proteins, play a significant role in processes such as cell adhesion, development, differentiation, and tumor progression [40]. Certainly the glycosynapse structure should be actively involved in the early events in host-pathogen interactions.

Lipid rafts are considered gateways in cellular membranes that allow communication between cells and play a role in the interaction between different organisms. Remarkably, glycosphingolipids or gangliosides are the receptors involved in the direct attachment to and invasion by some microbes and viruses of eukaryotic cells (Table 1). In some of these interactions, the sugar portion either in the microbe or in the ganglioside seems to be the molecular recognition target because exogenously added sugars interfere with some of these interactions. For instance, to adhere to epithelial cells, several pulmonary pathogenic bacteria recognize and bind to glycosphingolipids, and common saccharide moieties abrogate their adherence *in vitro* [57, 120]. Polysaccharides in a multivalent platform (dendrimers) seem to be a promising therapeutic tool in the control of the attachment of pathogens to host cells when gangliosides are involved [99].

Table 1. Microorganisms and viruses able to interact with lipid rafts, ceramides, or gangliosides

Microorganism or virus	Element recognized	Cell line employed	Reference
Viruses			
Avian sarcoma and leukosis virus	Entry requires lipid rafts	293 (human embryonic kidney)	[74]
Bovine adeno-associated virus	Glycosphingolipids are required for entry and transduction	COS (simian kidney), C6 (rat glioma), CHO mutants (Chinese hamster ovary)	[100]
BK virus	Infection requires gangliosides GT1b and GD1b and cholesterol	LNCaP (human prostate cancer), HPTE (human kidney proximal tubule epithelial), Vero (green monkey kidney)	[64, 27]
Coxsackievirus A9	Lipid raft integrity is required for infection	GMK (green monkey kidney), human monocytes	[123]
Coxsackievirus B4	Lipid raft integrity is required for infection	PANC-1 (human pancreas), ASPC-1 (human pancreas)	[124]
Ebola and Marburg viruses (EBOV/MBGV)	Lipid raft integrity and caveolae are required for infection	Peripheral Blood Mononuclear cells (PBMC), HEPG2 (human hepatocellular carcinoma), Vero-E6 (green monkey kidney), 293T (human fibroblast), HeLa (human cervical epithelia)	[9, 28]
Echovirus 1	Infection is dependent on cholesterol and caveosomes	GMK (green monkey kidney), CV-1 (green monkey kidney)	[90]
Echovirus 11	Entry is dependent on lipid rafts	HT29 (human colon cancer) RD (human rhabdomyosarcoma)	[115]
Human coronavirus 229E	Binds to CD13 in lipid microdomains, then clusters CD13 by cross-linking and uses caveolae to enter	Human skin fibroblast explants, L132 (human embryonic lung)	[79]
Human immunodeficiency virus type I (HIV-1)	Infection requires cholesterol, and fusion/entry steps may require lipid raft assembly	Jurkat (human T leukemia), PM1 (human cutaneous T-cell lymphoma from peripheral blood of a patient with Sezary Syndrome), CEM x 174 (human lymphocytes), BMVEC (human brain microvascular endothelia) CAEC (human coronary artery endothelia), HUVEC (human umbilical vein endothelia)	[61, 91, 63]
Human papilloma virus 31	Caveolae-dependent entry	HaCaT (human skin spontaneously immortalized), HEK-293T (human embryonic kidney SV40 T large antigen immortalized), low passage human foreskin keratinocytes	[112]

Table 1. (Continued)

Microorganism or virus	Element recognized	Cell line employed	Reference
Human rhinovirus	Attachment and entry induce formation of ceramide-enriched membrane-platforms and large glycosphingolipid-enriched membrane domains	HeLa (human cervical epithelia), Chang (human epithelial cells), HeLa (human cervical epithelia), WI-38 (human embryonic fibroblast)	[24, 36]
Human rotavirus	Binds GA2 glycolipid; GM1 involved in infection	MA104 (simian kidney)	[113, 38]
Human T cell leukemia virus type I (HTLV-I)	Cholesterol is required for entry	HOS (human fibroblast), 293T (human fibroblast), B5 (monkey lung fibroblast), and MAGI-5 (HeLa cells expressing CD4, CXCR4 and CCR5)	[134]
Influenza virus	Internalization via caveolae	MDCK (Darby canine kidney)	[80]
JC virus	Binds to lactosylceramide, and gangliosides: GM3, GD2, GD3, GD1b, GT1b, and GQ1b; Neoglycoprotein containing the terminal alpha2-6-linked sialic acid and GT1b inhibit infection.	IMR-32 (JCRB 9050; human neuroblastoma), SVG-A (human glial)	[56]
Murine hepatitis virus	Entry requires lipid rafts and cholesterol	DBT (mouse astrocytoma), 293A (human embryonic kidney); HeLa (human cervical epithelia), BHK (baby hamster kidney)	[17, 121]
Amphotropic murine leukemia virus	Entry route is caveola-dependent	NIH 3T3 (mouse embryo fibroblast), 293T (human fibroblast), CHO.K1 (Chinese hamster ovary), MDTF (*Mus dunni* tail fibroblastic)	[10]
Newcastle disease virus	Cholesterol, lipid rafts, and caveolae are required for entry	COS-7 (simian kidney), HeLa (human cervical epithelia), Vero (green monkey kidney)	[15]
Polyoma virus	Cholesterol, lipid rafts, caveolae, and gangliosides GD1a and GT1b are required for infection	C6 (rat glioma), human erythrocytes, Sf9 (*Spodoptera frugiperda*), NIH 3T6 (Swiss albino mouse), NMuMG (mouse mammary gland epithelia), CV-1 (monkey kidney epithelial)	[34, 126, 94]
Rabies virus	Gangliosides GD1b, GT1b and GQ1b are required for infectivity	BHK-21 (baby hamster kidney), CER (chick embryo-related)	[118]

Microorganism or virus	Element recognized	Cell line employed	Reference
Respiratory syncytial virus (RSV)	Uptake is dependent on caveolae	Cattle and human dendritic cells differentiated from peripheral blood mononuclear cells.	[133]
Semliki forest virus	Cholesterol is required for infection	BHK-21 (baby hamster kidney) CHO (Chinese hamster ovary) C6/C3 (clonal cell line derived from *Aedes albopictus*)	[89]
Sendai virus	Binds ganglioside GD1a	CV-1 (monkey kidney fibroblast)	[95]
Simian virus 40	Virus receptor ganglioside GM1	C6 (rat glioma), human erythrocytes, CV-1 (monkey kidney epithelial)	[126, 114]
Sindbis virus	Entry requires cholesterol	C6/C3 (clonal cell line derived from *Aedes albopictus*)	[65]
Vaccinia virus	Entry involves MHC class I molecule and caveolae pathway	BSC40 (green monkey kidney epithelia), HeLa (human cervical epithelia), HT-29 (human colon cancer), $LMtk^-$ (murine fibroblast mutants)	[18]
Bacteria			
Afipia felis	Integrity of lipid microdomains is required for entry	J774E (murine macrophage-like), RAW 264.7 (murine macrophage-like)	[102]
Anaplasma phagocytophilum	Entry and intracellular infection involve cholesterol-rich lipid rafts or caveolae and glycosylphosphatidylinositol (GPI)-anchored proteins	HL-60 (human leukemia)	[62]
Brucella abortus	Lipid rafts are involved in entry	Bone marrow-derived macrophages	[52]
Brucella spp.	Lipid rafts are involved in entry	J774.1 (murine macrophage), murine bone marrow derived macrophages	[75, 132]
Chlamydia spp.	Entry requires lipid rafts, cholesterol and caveolae	HeLa 229 (human cervical epithelia)	[116]
Campylobacter jejuni	Caveolae is involved in entry	INT407 (human embryonic intestinal) Caco-2 (human colon cancer)	[44, 135]
Escherichia coli ($FimH^+$)	Caveolae is involved in entry	Murine bone marrow-derived mast cells	[108]
Escherichia coli (Dr-positive)	Cholesterol is required for entry	HeLa (human cervical epithelia), Caco-2 (human colon cancer)	[49]

Table 1. (Continued)

Microorganism or virus	Element recognized	Cell line employed	Reference
Escherichia coli K1	Internalization is dependent on caveolae	HBMEC (human brain microvascular endothelia)	[117]
Escherichia coli (Type 1 Fimbria)	Invasion requires lipid rafts	5637 (human bladder epithelia)	[26]
Ehrlichia chaffeensis	Entry and intracellular infection involve cholesterol-rich lipid rafts or caveolae and glycosylphosphatidylinositol (GPI)-anchored proteins	THP-1 (human leukemia)	[62]
Haemophilus influenzae	Sialyl-lactosylceramide (GM3) is the minimal structure for fimbria-dependent binding, and gangliosides GM1, GM2, GM3, and GD1a are inhibitors of binding	Human erythrocytes and oropharyngeal epithelium	[128]
Legionella pneumophila	Cholesterol and lipid rafts are involved in entry	Murine-derived bone marrow macrophages	[7]
Listeria monocytogenes	Lipid rafts are involved in entry	Vero (African green monkey kidney epithelium), L2071 (mouse fibroblasts), Lovo (human epithelium), Rov-9 (rabbit kidney epithelium)	[105]
Moraxella catarrhalis	Asialo-GM1 and asialo-GM2 are involved in adherence	Human pharyngeal epithelial cells	[4]
Mycobacterium	Cholesterol is involved in entry	J744A.1 (murine macrophages), murine bone marrow-derived macrophages	[32]
Mycobacterium kansasii	Cholesterol is involved in entry	Human neutrophiles isolated from blood	[88]
Porphyromonas gingivalis	ICAM-1 and caveolae are required for invasion	KB (human oral epidermoid carcinoma)	[119]
Neisseriae gonorrhoeae	Cholesterol- and caveolae-dependent entry	293T (human embryonic kidney epithelia), primary human granulocytes isolated from fresh drawn blood	[101]
Pseudomonas aeuroginosa	Infection triggers activation of the acid sphingomyelinase and the release of ceramide in sphingolipid-rich rafts	Chang (conjunctiva epithelium), WI-38 (human fetal lung fibroblast), fibroblast, human nasal epithelium, murine lung epithelium, MDCK type II (Darby canine kidney)	[35, 21]

Microorganism or virus	Element recognized	Cell line employed	Reference
Salmonella typhimurium	Uptake is dependent on cholesterol	HeLa (human cervical epithelia), NIH3T3 (mouse embryo fibroblast)	[31]
Shigella flexneri	Lipid rafts are involved in binding and entry	HeLa (human cervical epithelia), CHO.K1 (Chinese hamster ovary)	[58]
Staphylococcus aureus	Recognition of asialo-GM1 initiate epithelial inflammatory response	1HAEo (SV40 immortalized human airway epithelial cells), CHO-K1 (Chinese hamster ovary)	[93]
Streptococcus gordonii (group A)	Caveolae is involved in invasion	Hep-2 (human larynx carcinoma), HUVEC (human umbilical vein endothelial)	[96]
Streptococcus pneumoniae	Adherence is dependent on lacto-N-neotetraose or asialoIganglioside GM1	Chinchilla tracheal epithelium	[122]
Streptococcus uberis	Caveolae is involved in entry	MAC-T (bovine mammary epithelia)	[6]
Vibrio cholerae	GM1 ganglioside recognition is required for transcytosis	Caco-2 (human colon cancer)	[11]
Yersinia enterocolitica	Intracellular growth requires lipid raft-mediated infection	J774.1 (murine macrophage), THP-1 (human monocytic)	[97]
Parasites			
Plasmodium falciparum	Host sphingomyelin-, cholesterol-, and lipid raft-associated proteins are internalized by the parasite	Human erythrocytes	[59]
Trypanosoma cruzi	Microdomains enriched in cholesterol and GM1 are involved in invasion	Vero (green monkey kidney) HeLa (human cervical epithelia)	[29]

Table 2. Toxins that recognize lipid rafts, glycolipids or gangliosides

Toxin/organism	Element recognized	Cell or tissue targeted	Reference
Viral origin			
Enterotoxin NSP4/Rotavirus	Caveolin-1	MDCK (Darby canine kidney), Caco-2 (human colon cancer), BSC-1 (African green monkey kidney)	[85]
Bacterial origin			
Aerolysin/*Aeromonas hydrophila*	Glycosyl phosphatidylinositol-anchored proteins and lipid rafts	BHK (baby hamster kidney)	[1]
Anthrax toxin/*Bacillus anthracis*	Lipid rafts	CHO (Chinese hamster ovary)	[2]
Botulinum toxin/*Clostridium* spp.	Complex gangliosides	Whole wild type mouse versus complex ganglioside deficient mouse	[53]
Cholera toxin/*Vibrio cholerae*	GM1, lipid rafts, and sucrose-isomaltose	Swine small intestine explants	[41]
Cry1A toxin/*Bacillus thuringiensis*	Insect-derived lipid rafts	Brush border membrane vesicles (insect-derived)	[136]
Cytolethal-distending toxin/*Actinobacillus actinomycetemcomitans*	Lipid rafts and cholesterol	Jurkat (T cell leukaemia)	[13]
Delta toxin/*Clostridium perfringens*	Ganglioside GM2	Human, goat, and guinea pig platelets	[47]
Epsilon toxin/*Clostridium perfringens*	Lipid rafts	mpkCCD$_{c14}$ (aldosterone-responsive kidney cortical collecting duct)	[16]
Exotoxin A/Pseudomonas	Lipid rafts	Vero (green monkey kidney), HeLa (human cervical epithelia), 3T3 (mouse embryo fibroblast)	[110]
Heat-labile enterotoxin (LT)/ Enterotoxigenic *Escherichia coli*	Lipid rafts	HT-29 (human colon cancer)	[51]
Heat-labile enterotoxin LTIIb/*Escherichia coli*	Ganglioside GD1a, Toll-like receptor II (TLR2)	CHO-K1 (Chinese hamster ovary), HEK-293 (human embryonic kidney), COS-7 (simian kidney), human blood derived monocytes, THP-1 (human monocytic), murine macrophages	[60]

Toxin/organism	Element recognized	Cell or tissue targeted	Reference
Hemolysin/*Vibrio cholerae*	Cholesterol	Rabbit erythrocytes, liposomes	[45]
Iota toxin/*Clostridium perfringens*	Lipid rafts	MDCK (Darby canine kidney)	[72]
Leukotoxin/*Actinobacillus actinomycatemcomitans*	Lipid rafts	Jn.9 (Jurkat human T leukemia expressing LFA-1), Jβ2.7 (LFA-1 deletion mutant)	[30]
Leukotoxin/*Mannheimia haemolytica*	Lipid rafts	BL-3 (bovine lymphoblastoid)	[8]
Lipoteichoic acid/Gram positive bacteria	Lipid rafts	Stable transfected HEK293 (human embryonic kidney) and Chinese hamster ovary	[125]
Listeriolysin O/*Listeria monocytogenes*	Lipid rafts and caveolae	HT29-MTX (mucus-secreting HT-29 from human colon cancer), Caco-2/TC7 (Caco-2 clone human colon cancer), J774 (murine macrophage-like)	[20, 33]
LPS/Gram negative bacteria	Lipid rafts	RAW 264.7 (murine macrophage-like)	[83]
Neurotoxin type B/*Clostridium botulinum*	Synaptotagmins and GT1b and GD1a gangliosides	Lipid vesicles	[78]
Neurotoxin type C/*Clostridium botulinum*	Caveolin-1	HT-29 (human colon cancer)	[127]
Perfringolysin O/*Clostridium perfringens*	Lipid rafts and cholesterol	MOLT-4 (Human acute lymphoblastic leukemia), human platelets and erythrocytes, A431 (human epidermal carcinoma)	[107, 130, 82]
Pertussis toxin/*Bordetella pertussis*	Lactosylceramide and gangliosides	Rabbits ciliated and macrophages cells	[98]
Shiga toxin/*Shigella dysenteriae*	Glycolipid globotriaosylceramide (Gb3)	HeLa (human cervical epithelia)	[46]
Shiga-like toxin-1/Enterohemorrhagic *Escherichia coli*	Glycolipid globotriaosylceramide (Gb3) and lipid rafts	Hela (human cervical epithelia), L3/bcl-2 (), Vero (green monkey kidney)	[111]
Streptolysin O/*Streptococcus* spp.	Cholesterol	Rabbit erythrocytes	[92]

Table 2. (Continued)

Toxin/organism	Element recognized	Cell or tissue targeted	Reference
Toxin A/*Clostridium difficile*	Pentasaccharide-ceramide (Gal alpha 1-3Gal beta 1-4GlcNAc beta 1-3Gal beta 1-4Glc-Cer) and decasaccharide-ceramide (Gal alpha 1-3Gal beta 1-4GlcNAc beta 1-3[Gal alpha 1-3Gal beta 1-4GlcNAc beta 1-6]Gal beta 1-4GlcNAc beta 1-3Gal beta 1-4Glc-Cer)	Rabbit erythrocytes	[19]
Tetanus toxin/*Clostridium tetani*	Lipid rafts, complex gangliosides, GD1b, and cholesterol	PC12 (), murine spinal cord cells, rat spinal cord motor neurons, whole wild type mouse versus complex gangliosides deficient mouse	[43, 23, 54, 53]
Vacuolating toxin (VacA)/*Helicobacter pylori*	Lipid rafts	HeLa (human cervical epithelia), CHO (Chinese hamster ovary),	[103]
Verotoxin 1/*Escherichia coli*	Glycolipid globotriosylceramide	Daudi (human lymphoma), HFB18.5 (human fetal B-cell)	[129]
Eukaryotic origin			
Granulysin/Cytotoxic T lymphocytes	Lipid rafts	Human dendritic cells isolated from blood	[131]
Sticholysin II toxin (actinoporin family)/*Stichodactyla helianthus*	Lipid rafts	COS-7 (simian kidney)	[5]

The degree to which the recognition of saccharides and gangliosides in the host-pathogens interactions is widespread and significant needs further analysis and should be a research topic of high priority.

Gangliosides and Lipid Raft Participation in Toxin-Targeted Cell Interactions

Numerous toxins of bacterial, viral or eukaryotic origin are able to interact with gangliosides, lipid raft located proteins, cholesterol, or lipid rafts directly to invade or intoxicate target cells as enumerated in Table 2. Some toxins use the cellular gateways to interact with eukaryotic cells and to gather valuable nutrients from dying cells for the benefit of the microbe. Eventually the toxins destroy the cells they have entered. However, there is a toxin of eukaryotic origin that uses lipid rafts to enter target cells but with the objective to protect the organism. This is granulysin—secreted by CD8-positive T lymphocytes—which enters infected dendritic cells through lipid rafts to kill intracellular bacteria [131]. Uptake into CHO cells of another special protein, the prion—that will lead to fatal neurodegenerative diseases—is also through a caveolae-mediated mechanism [87]. Interestingly, several toxins, such as the cholera toxin (CT) and the heat-labile toxin (LT), have outstanding properties as experimental mucosal adjuvants. Thus, these toxins, given in conjunction with antigens, are able to induce specific immune responses probably because they are inducing the recruitment and sampling activity of antigen-presenting cells that ultimately will be involved in the induction of immune responses. Another interesting application of a ganglioside-recognizing toxin is the delta toxin from *Clostridium perfringens*, which recognizes the ganglioside GM2 present in certain tumoral cells; remarkably this toxin has been used as an anticancer agent [47].

Interestingly, the ability of toxins to recognize specific glycolipids may determine their tissue specificity and ultimately the outcome of infection. For example, verotoxins (also known as shiga-like toxins) secreted by *Escherichia coli* O157:H7 bacteria recognize globo-series glycolipids and are associated with human enterohemorrhagic disease. However, changing the glycolipids recognized *in vitro* from Gb3 to Gb4 can produce a toxin able to mimic the human pathology in pigs [14]. The specific cytotoxic activity of verotoxin against the Vero cell line was neutralized *in vitro* by specific carbohydrates whose structure was designed after crystallographic analysis of the toxin [55]. These findings emphasize the key role of ganglioside recognition in host-pathogen interactions that can lead to dreadful diseases and they also indicate that saccharide moieties are the molecular targets involved. Overall, these findings are promising for the field of new therapeutics development, but they will need further *in vivo* treatment studies for determination of efficacy, tolerability, dosage, and more.

Gangliosides' Role in the Immune System

In the mucosal tissues any pathogen or antigen needs to cross the epithelial barrier in order to reach the antigen-presenting cells. The hallmark antigen-sampling cells in the epithelium are M cells. In my former work I determined that the *in vitro* sampling activity of human M-like cells of *V. cholerae* requires the recognition of GM1 [11]. Also, I showed that the pathogen-sampling activity of M-like cells is enhanced by the presence of opsonizing antibodies—secretory IgA derived from colostrums and IgG from serum—that recognize immobilized GM1 as well [12]. These findings showed that human antibodies' (secretory IgA or IgG) recognition of GM1 gangliosides has a functional implication in the sampling activity (internalization and transport) of *V. cholerae* through M-like cells [12]. More recently, a role for lipid rafts in the interaction of *V. cholerae* with M-like cells is supported by cholesterol depletion experiments I developed formerly. Depletion of cholesterol from the M-like cell membranes using methyl-β-cyclodextrin (MβCD) inhibits transcytosis of the bacteria through M-like cells in concentrations ranging from 0.5 mM up to 2.0 mM MβCD (Table 3). At 10 mM MβCD, a substantial disruption of the epithelial monolayer takes place (transepithelial resistance less than 120 Ohms x cm^2), allowing a higher bacterial transcytosis concomitantly (Table 3). At 5 mM MβCD, the transepithelial resistance is reduced in comparison to cells not treated with MβCD, and bacterial transcytosis is enhanced in comparison to a concentration at 2mM MβCD. This suggests that membrane integrity is damaged, leading to the higher transcytosis level. These results show that lipid raft domains play a role in the pathogen-sampling activity of M-like cells, and the results will need further *in vivo* confirmation.

Table 3. Effect of MβCD on the transcytosis of *V. cholerae* and the integrity of the tight junctions

Measurement at 2 hours	Methyl-beta-cyclodextrin concentration					
	None	0.5 mM	1.0 mM	2.0 mM	5.0 mM	10.0 mM
Real Transcytosis (%)	1.190 ± 0.780 (n=8)	0.628 ± 0.249 (n=4)	0.273 ± 0.160 (n=8)	0.166 ± 0.073* (n=6)	0.505 ± 0.102 (n=3)	16.278 ± 0.390 (n=3)
Transepithelial resistance (Ohms x cm^2)	229.4 ± 39.1	242.8 ± 40.5	236.3 ± 43.5	256.7 ± 41.8	132.7 ± 6.7	78.0 ± 26.9

The M-like cells were treated for 30 minutes with the indicated amount of MβCD in DMEM devoid of fetal serum and were then infected with *V. cholerae*. Real transcytosis was measured, quantifying the number of colony forming units (cfu) in lumenal or basolateral supernatants at the time indicated, and calculated by applying the following formula: (basolateral cfu/ basolateral + lumenal cfu) x 100. The average + SEM of at least three different experiments are shown. *p< 0.5, Mann Whitney test.

The next process after antigen sampling that is also important in the induction of the immune response is the antigen presentation. Lipid rafts enriched in ganglioside GM1 participate also in the antigen presentation process because lipid raft platforms structure the immunological synapses where immune cells communicate to initiate cell activation and differentiation processes—both instrumental for the induction of the immune response [69]. Of interest as well, glycosphingolipids per se can act as antigens; NKT cells via CD1 molecules present gangliosides to autoreactive, multiple sclerosis patient-derived T cells [73]. In addition, adjuvants such as CT and LT (hallmark experimental adjuvants) enhance the immune response against an antigen. Both are bacteria-derived toxins that recognize gangliosides [41, 60]. In particular, heat-labile enterotoxin LT-IIa from *Escherichia coli* requires recognition of GM1 gangliosides to be able to induce immune responses in mice, following nasal vaccination with the surface adhesin protein AgI/II of *Streptococcus mutans* antigen [77]. The B subunit of LT-IIa requires that TLR-2 and gangliosides act as co-receptors for the activation of human mononuclear cells [39]. *In vitro* studies have shown that the B subunit of the LT-IIb enterotoxin, which is a potent mucosal adjuvant itself, first recognizes the ganglioside GD1a and then cooperatively recruits, binds, and activates the toll-like receptor-2 (TLR-2), using different cell lines [60]. If this mechanism can be generalized to any ganglioside ligand molecule, then these findings may provide insights into how the TLR pathway can be indirectly activated by first recognition of the gangliosides.

Another well-established experimental mucosal adjuvant is CT, from *V. cholerae*, which recognizes the ganglioside GM1. One of the most ubiquitous and abundant gangliosides in lipid rafts is GM1; in fact it has been widely used as a marker of lipid rafts in tissues of different species and in cell line research studies [86]. GM1 plays a role in the interaction of antigen-presenting dendritic cells with cholera toxin [37]; and NFkB translocation from the cytoplasm to the nuclei is induced in dendritic cells after CT recognition of the GM1 [50]. Interestingly, antigen bridged to GM1 in B lymphocytes leads to antigen-efficient processing and presentation that is delayed, compared to B cell receptor antigen internalization [76]. In addition, bridging antigens to GM1, using a recombinant epitope-B subunit of LT, can lead to the development of antigen presentation in the MHC class I context, which will be instrumental in the induction of the specific TCD8 immune response as well [22]. In total, these findings show that GM1 is involved in antigen presentation, a key process in the immune system.

Another aspect of the immune response is the effector phase, where elimination of the specific pathogen and infected tissues will be accomplished usually by the cellular immune arm of the immune system. In the cellular immune responses where macrophages and TCD8-positive cells are the main players, ganglioside GM1 plays a role too. Recently, a new class of human monocytes (precursors of macrophages) with a high-level expression of GM1 gangliosides has been described. This population of monocytes expresses higher levels of GM1, has enhanced phagocytosis rates, and is more resistant to LPS than monocytes expressing low levels of GM1 [71]. Also, high-level expression of GM1 in TCD8-positive cells is involved in clearing of the virus of respiratory syncytial virus (RSV)-infected mice [70]. During RSV infection, an endogenous sialidase is induced in the pulmonary tissues, leading to enhanced expression of GM1 in CD8+ T lymphocytes [70]. In the clearance of RSV, this high level of GM1 cell surface expression in CD8+ T cells seems to be critical, because oseltamivir (a sialidase/neurominidase inhibitor and also an anti-influenza drug that reduces the level of GM1 on CD8+ T lymphocytes) blocks viral clearance [70]. Overall the evidence presented here suggests that gangliosides—in particular GM1—have a significant role in cellular effectors immune responses.

Interestingly, some tumor cells secrete gangliosides. Knowing that gangliosides insert into membranes [104], this strategy may allow tumor cells to modify their interactions with other cells and to regulate the immune response [106].

Ganglioside Participation in Signal Transduction

An intriguing aspect about gangliosides and their role in the processes described above is how gangliosides recognized by a ligand will trigger such diverse cellular responses. One way to achieve this is through the initiation of signal transduction cascades that can be effectively triggered by TLR [84]. Recent evidence is beginning to emerge that indicates a role of TLR on the ganglioside signal transduction process. The first hint about the ganglioside-TLR connection came from studies using *Pseudomonas aeruginosa* flagellin, the protein that forms the flagellum and is a ligand for TLR5 [42]. Flagellin induces quick responses in certain airway epithelial cells, despite the fact that these cells initially lack lumenal surface expression of TLR5. However, *P. aeruginosa* flagellin also recognizes lumenal-expressed asialo GM1 in airway epithelial cells [93]. Subsequently, Adamo et al. showed that activation of airway epithelial cells requires concomitant recognition of asialo-GM1, TLR5, and TLR2, the expression of which will be induced after initial recognition of the ganglioside [3]. Another example linking gangliosides to TLR expression was reported this same year with flagellin FliC from *Salmonella enteritidis.* FliC induces the expression of human β-defensin-2 in the human epithelial cell line Caco-2 via TLR5 activation and ganglioside (in particular GD1a) binding [81]. Later, the cooperative relationship between asialo GM1 and TLR5 was further analyzed and found to be required in the activation of the Erk1/2 signaling pathway in response to flagellin [67]. Moreover, antibodies against asialo GM1 promote the secretion of IL-8 in human corneal epithelial cells by a molecular mechanism that requires co-expression and protein-protein interaction between TLR5 and the asialo GM1 gangliosides in the surface of the cells [25]. The ganglioside-TLR relationship also has been found in brain cell responses in both rat-derived astrocytes and microglia, where gangliosides induce proinflammatory cytokine secretion via the NFκB pathway that requires TLR4 participation [48]. Also, as previously described *in vitro* and using different cell lines, the LT-IIb toxin and mucosal adjuvant first recognizes ganglioside GD1a and then cooperatively recruits, binds, and activates TLR-2 [60]. These studies suggest an important role for gangliosides during TLR activation and initiations of the immune response against pathogens, and, of course, further studies are needed to explore this relationship extensively.

Conclusion

Overall this chapter emphasizes the importance of gangliosides in a variety of interactions, including host-pathogens, toxin-target cells, and processes such as antigen sampling, antigen presentation, and cellular immune effector responses key to the immune system. Interestingly TLRs are emerging as partners in the molecular mechanism of signal transduction triggered by ganglioside ligands, and saccharide moieties recognition seems to be one important molecular target. Understanding the molecular mechanisms involved in the host-pathogen interaction will aid us to manipulate and to develop new tools aimed at modifying the early steps as well as the final outcomes in these diverse and critical interactions.

References

[1] Abrami, L. and van der Goot, G.F. (1999). Plasma membrane microdomains act as concentration platforms to facilitate intoxication by aerolysin. *J. Cell. Biol*, 147, 175-184.

[2] Abrami, L.; Liu, S.; Cosson, P.; Leppla, S.H. and Gisou van der Goot, F. (2003). Anthrax toxin triggers endocytosis of its receptor via a lipid raft-mediated clathrin-dependent process. *J. Cell Biol*, 160, 321-328.

[3] Adamo, R.; Sokol, S.; Soong, G.; Gomez, M.I. and Prince, A. (2004). *Pseudomonas aeruginosa* flagella activate airway epithelial cells through asialoGM1 and toll-like receptor 2 as well as toll-like receptor 5. *Am. J. Respir, Cell Mol. Biol*, 30, 627-634.

[4] Ahmed, K.; Suzuki, Y.; Miyamoto, D. and Nagatake, T. (2002). Asialo-GM_1 and asialo-GM_2 are putative adhesion molecules for *Moraxella catarrhalis. Med. Microbiol., Immunol.*, 191, 5-10.

[5] Alegre-Cebollada, J.; Rodríguez-Crespo, I.; Gavilanes, J.G. and Martinez del Pozo, A. (2006). Detergent-resistant membranes are platforms for actinoporin pore-forming activity on intact cells. *FEBS Journal*, 273, 863-871.

[6] Almeida, R.A. and Oliver, S.P. (2006). Trafficking of *Streptococcus uberis* in bovine mammary epithelial cells. *Microb. Pathog.*, 41, 80-89.

[7] Amer, A.O.; Byrne, B.G. and Swanson, M.S. (2005). Macrophages rapidly transfer pathogens from lipid raft vacuoles to autophagosomes. *Autophagy*, 1, 53-58.

[8] Atapattu, D.N. and Czuprynski C.J. (2007). *Mannheimia haemolytica* leukotoxin binds to lipid rafts in bovine lymphoblastoid cells (BL-3) and is internalized in a dynamin-2 and clathrin-dependent manner. *Infect. Immun.*, doi:10.1128/IAI.00534-07.

[9] Bavari, S.; Bosio, C.M.; Wiegand, E.; Ruthel, G.; Will, A.B.; Geisbert, T.W.; Hevey, M.; Schmaljohn, C.; Schmaljohn, A. and Aman, M.J. (2002). Lipid raft microdomains: a gateway for compartmentalized trafficking of Ebola and Marburg viruses. *J. Exp. Med.*, 195, 593-602.

[10] Beer, C.; Andersen, D.S.; Rojek, A. and Pedersen, L. (2005). Caveola-dependent endocytic entry of amphotropic murine leukemia virus. *J. Virol.*, 79, 10776-10787.

[11] Blanco, L.P. and DiRita, V.J. (2006). Bacterial-associated cholera toxin and GM_1 binding are required for transcytosis of classical biotype *Vibrio cholerae* through an *in vitro* M cell model system. *Cell. Microbiol.*, 8, 982-998.

[12] Blanco, L.P. and DiRita, V.J. (2006). Antibodies enhance interaction of *Vibrio cholerae* with intestinal M-like cells. *Infect. Immun.*, 74, 6957-6964.

[13] Boesze-Battaglia, K.; Besack, D.; Jordan-Sciutto, K. and Shenker, B.J. (2006). Cholesterol-rich membrane microdomains mediate cell cycle arrest induced by *Actinobacillus actinomycetemcomitans* cytolethal-distending toxin. *Cell. Microbiol.*, 8, 823-836.

[14] Boyd, B.; Tyrrell, G.; Maloney, M.; Gyles, C.; Brunton, J. and Lingwood, C. (1973). Alteration of the glycolipid binding specificity of the pig edema toxin from globotetraosyl to globotriaosyl ceramide alters in vivo tissue targeting and results in a verotoxin1-like disease in pigs. *J. Exp. Med.,* 177, 1745-1753.

[15] Cantín, C.; Holguera, J.; Ferreira, L.; Villar, E. and Muñoz-Barroso, I. (2007). Newcastle disease virus may enter cells by caveolae-mediated endocytosis. *J. Gen. Virol.*, 88, 559-569.

[16] Chassin, C.; Bens, M.; de Barry, J.; Courjaret, R.; Bossu, J.L.; Cluzeaud, F.; Ben Mkaddem, S.; Gilbert, M.; Poulain, B.; Popoff, M.R. and Vandewalle, A. (2007). Pore-forming epsilon toxin causes membrane permeabilization and rapid ATP depletion-mediated cell death in renal collecting duct cells. *Am. J. Physiol. Renal Physiol.* In press doi:10.1152/ajrenal.00199.2007.

[17] Choi, K.S.; Aizaki, H. and Lai, M.M. (2005). Murine Coronavirus requires lipid rafts for virus entry and cell-cell fusion but not for virus release. *J. Virol.*, 79, 9862-9871.

[18] Chung, C-S.; Huang, C-Y. and Chang, W. (2005). Vaccinia virus penetration requires cholesterol and results in specific viral envelope proteins associated with lipid rafts. *J. Virol.*, 79. 1623-1634.

[19] Clark, G.F.; Krivan, H.C.; Wilkins, T.D. and Smith, D.F. (1987). Toxin A from *Clostridium difficile* binds to rabbit erythrocyte glycolipids with terminal Gal alpha 1-3Gal beta 1-4GlcNac sequences. *Arch. Biochem. Biophys.*, 257, 217-229.

[20] Coconnier, M-H.; Lorrot, M.; Barbat, A.; Laboisse, C. and Servin, A. (2000). Listeriolysin O-induced stimulation of mucin exocytosis in polarized intestinal mucin-secreting cells: evidence for toxin recognition of membrane-associated lipids and subsequent toxin internalization through caveolae. *Cell. Microbiol.,* 2, 487-504.

[21] Comolli, J.C.; Waite, L.L.; Mostov, K.E.; and Engel, J.N. (1999). Pili binding to asialo-GM_1 on epithelial cells can mediate cytotoxicity or bacterial internalization by *Pseudomonas aeruginosa. Infect. Immun.,* 67, 3207-3214.

[22] De Haan, L.; Hearn, A.R.; Rivett, J.A. and Hirst, T.R. (2002). Enhanced delivery of exogenous peptides into class I antigen processing and presentation pathway. *Infect. Immun.*, 70, 3249-3258.

[23] Deinhardt, K.; Berninghausen, O.; Willison, H.J.; Hopkins, C.R. and Schiavo, G. (2006). Tetanus toxin is internalized by a sequential clathrin-dependent mechanism initiated within lipid microdomains and independent of epsin 1. *J. Cell Biol.*, 174, 459-471.

[24] Dreschers, S.; Franz, P.; Dumitri, C.A.; Wilker, B.; Jahnke, K. and Gulbins, E. (2007). Infections with human Rhinovirus induce the formation of distinct functional membrane domains. *Cell. Physiol. Biochem.*, 15, 241-254.

[25] Du, J.W.; Zhang, F.; Capó-Aponte, J.E.; Tachado, S.D.; Zhang, J.; Yu, F-S.X.; Sack, R.; Koziel, H. and Reinado, P.S. (2006). AsialoGM1-mediated IL-8 release by human

corneal epithelial cells requires coexpression of TLR5. *Invest. Ophtalmol. Visual Sci.*, 47, 4810-4818.

[26] Duncan, M.J.; Li, G.; Shin, J-S.; Carson, J.L. and Abraham, S.N. (2004). Bacterial penetration of bladder epithelium through lipid rafts. *J. Biol. Chem.*, 279, 18944-18951.

[27] Eash, S.; Querbes, W. and Atwood, W.J. (2004). Infection of Vero cells by BK virus is dependent on caveolae. *J. Virol.,* 78, 11583-11590.

[28] Empig, C.J and Goldsmith, M.A. (2002). Association of the caveola vesicular system with cellular entry of filoviruses. *J. Virol.,* 76, 5266-5270.

[29] Fernandes, M.C.; Cortez, M.; Aparecida, K.; Yoneyama, G.;, Straus, A.H.; Yoshida, N. and Mortara, R.A. (2007). Novel strategy in *Trypanosoma cruzi* cell invasion: implication of cholesterol and host cell microdomains. *Int. J. Parasitol,* doi:10.1016/j.ijpara.2007.04.025.

[30] Fong, K.P.; Pacheco, C.M.F.; Otis, L.L.; Baranwal, S.; Kieba, I.R.; Harrison, G.; Hersh, E.V.; Boesze-Battaglia and Lally, E.T. (2006). *Actinobacillus actinomycetemcomitans* leukotoxin requires lipid microdomains for target cell cytotoxicity. *Cell. Microbiol.*, 8, 1753-1767.

[31] Garner, M.J.; Hayward, R.D. and Koronakis, V. (2002). The *Salmonella* pathogenicity island 1 secretion system directs cellular cholesterol redistribution during mammalian cell entry and intracellular trafficking. *Cell Microbiol.*, 4, 153-165.

[32] Gatfield, J. and Pieters, J. (2000). Essential role for cholesterol in entry of Mycobacteria into macrophages. *Science*, 288, 1647-1650.

[33] Gekara, N.O.; Jacobs, T.; Chakraborty, T. and Weiss, S. (2005). The cholesterol-dependent cytolysin listeriolysin O aggregates rafts via oligomerization. *Cell. Microbiol.*, 7, 1345-1356.`

[34] Gilbert, J. and Benjamin, T. (2004). Uptake pathway of polyomavirus via ganglioside GD1a. *J. Virol.*, 78, 12259-12267.

[35] Grassmé, H.; Jendrossek, V.; Riehle, A.; von Kurthy, G.; Berger, J.; Schwarz, H.; Weller, M.; Kolesnick, R. and Gulbins, E. (2003). Host defense against *Pseudomonas aeruginosa* requires ceramide-rich membrane rafts. *Nature Medicine*, 9, 322-330.

[36] Grassmé, H.; Riehle, A.; Wilker, B. and Gulbins, E. (2005). Rhinoviruses infect human epithelial cells via ceramide-enriched membrane platforms. *J. Biol. Chem.*, 280, 26256-26262.

[37] Grdic, D.; Ekman, L.; Schön, K.; Lindgren, K.; Mattsson, J.; Magnusson, K-E.; Ricciardi-Castagnoli, P. and Lycke, N. (2005). Splenic marginal zone dendritic cells mediate the cholera toxin adjuvant effect: dependence on the ADP-ribosyltranferase activity of the holotoxin. *J. Immunol.*, 175, 5192-5202.

[38] Guo, C-T.; Nakagomi, O.; Mochizuki, M.; Ishida, H.; Kiso, M.; Ohta, Y.; Suzuki, T.; Miyamoto, D.;Jwa Hidari, K. I-P. and Suzuki, Y. (1999). Ganglioside GM_{1a} on the cell surface is involved in the infection by human rotavirus KUN and MO strains. *J. Biochem.*, 126, 683-688.

[39] Hajishengallis, G.; Tapping, R.I.; Martin, M.H.; Nawar, H.; Lyle, E.A.; Russell, M.W. and Connell, T.D. (2005). Toll-like receptor 2 mediates cellular activation by the B subunits of type II heat-labile enterotoxins. *Infect. Immun.*, 73, 1343-1349.

[40] Hakomori, S. (2002). The glycosynapse. *Proc. Natl. Acad. Sci. USA,* 99, 225-232.

[41] Hansen, G.H.; Dalskov, S-M.; Rehne Rasmussen, C.; Immerdal, L.; Niels-Christiansen, L-L. and Danielsen, M.E. (2005). Cholera toxin entry into pig enterocytes occurs via a lipid raft- and clathrin-dependent mechanism. *Biochem.*, 44, 873-882.

[42] Hayashi, F.; Smith, K.D.; Ozinsky, A.; Hawn, T.R.; Yi, E.C.; Goodlett, D.R.; Eng, J.K.; Akira, S.; Underhill, D.M. and Aderem, A. (2001). The innate immune response to bacterial flagellin is mediated by Toll-like receptor 5. *Nature*, 410, 1099-1103.

[43] Herreros, J.; Ng, T. and Schiavo, G. (2001). Lipid rafts act as specialized domains for tetanus toxin binding and internalization into neurons. *Mol. Biol. Cell*, 12, 2947-2960.

[44] Hu, L.; McDaniel, J.P. and Kopecko, D.J. (2006). Signal transduction events in human epithelial cell invasion by *Campylobacter jejuni* 81-176. *Microb. Pathog.,* 40, 91-100.

[45] Ikigai, H.; Akatsuka, A.; Tsujiyama, H.; Nakae. T. and Shimamura, T. (1996). Mechanism of membrane damage by El Tor hemolysin of *Vibrio cholerae* O1. *Infect. Immun.*, 64, 2968-2973.

[46] Jacewicz, M.; Clausen, H.; Nudelman, E.; Donohue-Rolfe, A. and Keusch, G.T. (1986). Pathogenesis of Shigella diarrhea XI. Isolation of a *Shigella* toxin-binding glycolipid from rabbit jejunum and HeLa cells and its identification as globotriaosylceramide. *J. Exp. Med.*, 163, 1391-1404.

[47] Jolivet-Reynaud, C.; Launay, J-M. and Alouf, J.E. (1988). Damaging effects of *Clostridium perfringens* Delta toxin on blood platelets and their relevance to ganglioside GM2. *Arch. Biochem. Biophys.,* 262, 59-66.

[48] Jou, I.; Lee, J.H.; Park, S.Y.; Yoon, H.J.; Joe, E-H. and Park, E.J. (2006). Gangliosides trigger inflammatory responses via TLR4 in brain glia. *Am. J. Pathol.,* 168, 1619-1630.

[49] Kansau, I.; Berger, C.; Hospital, M.; Amsellem, R.; Nicolas, V.; Servin, A.L. and Bernet-Camard, M-F. (2004). Zipper-like internalization of Dr-positive *Escherichia coli* by epithelial cells is preceded by an adhesion-induced mobilization of raft-associated molecules in the initial step of adhesion. *Infect. Immun*., 72, 3733-3742.

[50] Kawamura, Y.I.; Kawashima, R.; Shirai, Y.; Kato, R.; Hamabata, T.; Yamamoto, M.; Furukawa, K.; Fujihashi, K.; McGhee, J.R.; Hayashi, H. and Dohi, T. (2003). Cholera toxin activates dendritic cells through dependence on GM1-ganglioside which is mediated by NF-κB translocation. *Eur. J. Immunol.*, 33, 3205-3212.

[51] Kesty, N.C.; Mason, K.M.; Reedy, M.; Miller, S.E.; and Kuehn, M.J. (2004). Enterotoxigenic *Escherichia coli* vesicles target toxin delivery into mammalian cells. *EMBO J.,* 23, 4538-4549.

[52] Kim, S.; Watarai, M.; Suzuki, H.; Makino, S-i.; Kodama, T. and Shirahata, T. (2004). Lipid raft microdomains mediate class A scavenger receptor-dependent infection of *Brucella abortus. Microbial. Pathog.,* 37, 11-19.

[53] Kitamura, M.; Takamiya, K.; Aizawa, S.; Furukawa, K. and Furukawa, K. (1999). Gangliosides are the binding substances in neural cells for tetanus and botulinum toxins in mice. *Biochem. Biophys. Acta*, 1441, 1-3.

[54] Kitamura, M.; Igimi, S.; Furukawa, K. and Furukawa, K. (2005). Different response of the knockout mice lacking b-series gangliosides against botulinum and tetanus toxins. *Biochim. Biophys. Acta*, 1741, 1-3.

[55] Kitov, P.I.; Sadowska, J.M.; Mulvey, G.; Armstrong, G.D.; Ling, H.; Pannus, N.S.; Read, R.J. and Bundle, D.R. (2000). Shiga-like toxins are neutralized by tailored multivalent carbohydrates ligands. *Nature*, 403, 669-672.

[56] Komagome, R.; Sawa, H.; Suzuki, T.; Suzuki, Y.; Tanaka, S.; Atwood, W.J. and Nagashima, K. (2002). Oligosaccharides as receptors for JC virus. *J. Virol.,* 76, 12992-13000.

[57] Krivan, H.C.; Roberts, D.D. and Ginsburg, V. (1988). Many pulmonary pathogenic bacteria bind specifically to the carbohydrate sequence GalNAc beta 1-4Gal found in some glycolipids. *Proc. Natl. Acad. Sci. USA*, 85, 6157-6161.

[58] Lafont, F.; Tran Van Nhieu, G.; Hanada, K.; Sandonetti, P. and van der Goot F.G. (2002). Initial steps of *Shigella* infection depend on the cholesterol/sphingolipid raft-mediated CD44-IpaB interaction. *EMBO J.,* 21, 4449-4457.

[59] Lauer, S.; VanWye, J.; Harrison, T.; McManus, H.; Samuel, B.U.; Hiller, L.N.; Mohandas, N. and Haldar, K. (2000). Vacuolar uptake of host components, and a role for cholesterol and sphingomyelin in malarial infection. *EMBO J.,* 19, 3556-3564.

[60] Liang, S.; Wang, M.; Tapping, R.I.; Stepensky, V.; Nawar, H.F.; Triantafilou, M.; Triantafilou, K.; Connell, T.D. and Hajishengallis, G. (2007). Ganglioside GD1a is an essential coreceptor for toll-like receptor 2 signaling in response to the B subunit of type IIb enterotoxin. *J. Biol. Chem.*, 282, 7532-7542.

[61] Liao, Z.; Cimakasky, L.M.; Hampton, R.; Nguyen, D.H. and Hildreth, J.E.K. (2001). Lipid rafts and HIV pathogenesis: host membrane cholesterol is required for infection by HIV type I. *AIDS Res. Human. Retroviruses*, 17, 1009-1019.

[62] Lin, M. and Rikihisa, Y. (2003). Obligatory intracellular parasitism of *Erlichia chaffeensis* and *Anaplasma phagocytophilum* involves caveolae and glycosylphosphatidylinositol-anchored proteins. *Cell. Microb.*, 5, 809-820.

[63] Liu, N.Q.; Lossinsky, A.S.; Popik, W.; Li, X.; Gujuluva, C.; Kriederman, B.; Roberts, J.; Pushkarsky, T.; Bukrinsky, M.; Witte, M.; Weinand, M. and Fiala, M. (2002). Human immunodeficiency virus type 1 enters brain microvascular endothelia by macropinocytosis dependent on lipid rafts and mitogen-activates protein kinase signaling pathway. *J. Virol.*, 76, 6689-6700.

[64] Low, J.A.; Magnuson, B.; Tsai, B. and Imperiale, M.J. (2006). Identification of ganglioside GD1b and GT1b as receptors for BK virus. *J. Virol.*, 80, 1361-1366.

[65] Lu, Y.E.; Cassese, T. and Kielian, M. (1999). The cholesterol requirement for Sindbis virus entry and exit and characterization of a spike protein region involved in cholesterol dependence. *J. Virol.,* 73, 4272-4278.

[66] Mañez, S.; del Real, G. and Martínez-A, C. (2003). Pathogens: Raft hijackers. *Nature Rev. Immunol.,* 3, 557-568.

[67] McNamara, N.; Gallup, M.; Sucher, A.; Maltseva, I.; McKemy, D. and Basbaum, C. (2006). AsialoGM1 and TLR5 cooperate in flagellin-induced nucleotide signaling to activate Erk1/2. *Am. J. Respir. Cell Mo. Biol.*, 34, 653-660.

[68] Michel, V. and Bakovik, M. (2007). Lipid rafts in health and disease. *Biol. Cell*, 99, 129-140.

[69] Monks, C.R.; Freiberg, B.A.; Kupfer, H.; Sciaky, N. and Kupfer, A. (1998). Three-dimensional segregation of supramolecular activation clusters in T cells. *Nature*, 395, 82-86.

[70] Moore, M.L.; Chi, M.H.; Zhou, W.; Goleniewska, K.; O'Neal, J.F.; Higginbotham, J.N. and Peebles Jr., R.S. (2007). Cutting Edge: Oseltamivir decreases T cell GM1 expression and inhibits clearance of respiratory syncytial virus: potential role of endogenous sialidase in antiviral immunity. *J. Immunol.,* 178, 2651-2654.

[71] Moreno-Altamirano, M.M.B.; Aguilar-Carmona, I. and Sánchez-García, F.J. (2007). Expression of GM1, a marker of lipid rafts, defines two subsets of human monocytes with differential endocytic capacity and lipopolysaccharide responsiveness. *Immunol.*, 120, 536-543.

[72] Nagahama, M.; Yamaguchi, A.; Hagiyama, T.; Ohkubo, N.; Kobayashi, K. and Sakurai, J. (2004). Binding and internalization of *Clostridium perfringens* iota-toxin in lipid rafts. *Infect. Immun.,* 72, 3267-3275.

[73] Naidenko, O.L.; Koezuka, Y. and Kronenberg, M. (2000). CD1-mediated antigen presentation of glycosphingolipids. *Microb. Infect.*, 2, 621-631.

[74] Narayan, S.; Barnard, R.J.O. and Young, J.A.T. (2003). Two retroviral entry pathways distinguished by lipid raft association of the viral receptor and differences in viral infectivity. *J. Virol.,* 77, 1977-1983.

[75] Naroeni, A. and Porte, F. (2002). Role of cholesterol and the ganglioside GM_1 in entry and short-term survival of *Brusella suis* in murine macrophages. *Infect. Immun.*, 70, 1640-1644.

[76] Nashar, T.O.; Betteridge, Z.E. and Mitchell, R.N. (2002). Antigen binding to GM_1 ganglioside results in delayed presentation: minimal effects of GM_1 on presentation of antigens internalized via other pathways. *Immunol.,* 106, 60-70.

[77] Nawar, H.F.; Arce, S.; Russell, M.W. and Connell, T.D. (2007). Mutants of type II heat-labile enterotoxin LT-IIa with altered ganglioside-binding activities and diminished toxicity are potent mucosal adjuvants. *Infect. Immun*., 75, 621-633.

[78] Nishiki, T.; Tokuyama, Y.; Kamata, Y.; Nemoto, Y.; Yoshida, A.; Sato. K.; Sekiguchi, M.; Takahashi, M. and Kozaki, S. (1996). The high-affinity binding of *Clostridium botulinum* type B neurotoxin to synaptotagmin II associated with gangliosides GT1b/GD1a. *FEBS Letters*, 378, 253-257.

[79] Nomura, R.; Kiyota, A.; Suzaki, E.; Kataoka, K.; Ohe, Y.; Miyamoto, K.; Senda, T. and Fujimoto, T. (2004). Human coronavirus 229E binds in rafts to CD13 and enters the cell through caveolae. *J. Virol*., 78, 8701-8708.

[80] Nunes-Correia, I.; Eulálio, A.; Nir, S. and Pedroso de Lima, M.C. (2004). Caveolae as an additional route for influenza virus endocytosis in MDCK cells. *Cell. Molec. Biol. Lett.,* 9, 47-60.

[81] Ogushi, K-i; Wada, A.; Niidome, T.; Okuda, T.; Llanes, R.; Nakayama, M.; Nishi, Y.; Kurazono, H.; Smith, K.D.; Aderem, A.; Moss, J. and Hirayama, T. (2004). Gangliosides act as co-receptors for *Salmonella enteritidis* FliC and promote FliC induction of human beta-defensin-2 expression in Caco-2 cells. *J. Biol. Chem.,* 279, 12213-12219.

[82] Ohno-Iwashita, Y.; Shimada, Y.; Waheed, A.A.; Hayashi, M.; Inomata, M.; Nakamura, M.; Maruya and Iwashita, S. (2004). Perfringolysin O, a cholesterol-binding cytolisin, as a probe for lipid rafts. *Anaerobe*, 10, 125-134.

[83] Olsson, S. and Sundler, R. (2006). The role of lipid rafts in LPS-induced signaling in a macrophage cell line. *Mol. Immunol.*, 43, 607-612.

[84] O'Neill, L.A.J. and Bowie, A.G. (2007). The family of five: TIR-domain-containing adaptors in Toll-like receptor signaling. *Nature Rev. Immunol.,* 7, 353-364.

[85] Parr, R.D.; Storey, S.M.; Mitchell, D.M.; McIntosh, A.L.; Zhou, M.; Mir, K.D. and Ball, J. M. (2006). The rotavirus enterotoxin NSP4 directly interacts with the caveolar structural protein caveolin-1. *J. Virol.,* 80, 2842-2854.

[86] Parton, R.G. (1994). Ultrastructural localization of gangliosides; GM1 is concentrated in caveolae. *J. Histochem. Cytochem.*, 42, 155-166.

[87] Peters, P.J.; Mironov Jr., A.; Peretz, D.; van Donselaar, E.; Leclerc, E.; Erpel, S.; DeArmond, S.; Burton, D.R.; Williamson, R.A.; Vey, M. and Prusiner, S.B. (2003). Trafficking of prion proteins through a caveolae-mediated endosomal pathway. *J. Cell Biol*, 162, 703-717.

[88] Peyron, P.; Bordier, C.; N'Diaye, E-N. and Maridonneau-Parini, I. (2000). Nonopsonic phagocytosis of *Mycobacterium kansasii* by human neutrophiles depends on cholesterol and is mediated by CR3 associated with glycosylphosphatidylinositol-anchored proteins. *J. Immunol.,* 165, 5186-5191.

[89] Phalen, T. and Kielian, M. (1991). Cholesterol is required for infection by Semliki Forest virus. *J. Cell Biol.*, 112, 615-623.

[90] Pietiäinen, V.; Marjomaki, V.; Upla, P.; Pelkmans, L.; Helenius, A. and Hyypia, T. (2004). Echovirus 1 endocytosis into caveosomes requires lipid rafts, dynamin II, and signaling events. *Mol. Biol. Cell*, 15, 4911-4925.

[91] Popik, W.; Alce, T.M. and Au, W-C. (2002). Human immunodeficiency virus type 1 uses lipid raft-cololized CD4 and chemokine receptors for productive entry into CD4+ T cells. *J. Virol.,* 76, 4709-4722.

[92] Prigent, D. and Alouf, J.E. (1976). Interaction of Streptolysin O with sterols. *Biochim. Biophys. Acta*, 443, 288-300.

[93] Ratner, A.J.; Bryan, R.; Weber, A.; Nguyen, S.; Barnes, D.; Pitt, A.; Gelber, S.; Cheung, A. and Prince, A. (2001). Cystic fibrosis pathogens activate Ca^{2+}-dependent mitogen-activated protein kinase signaling pathways in airway epithelial cells. *J. Biol. Chem.*, 276, 19267-19275.

[94] Richterová, Z.; Liebl, D.; Horák, M.; Palková, Z.; Stokrová, J.; Hozák, P.; Korb, J. and Forstová, J. (2001). Caveolae are involved in the trafficking of mouse polyomavirus virions and artificial VP1 pseudocapsids toward cell nuclei. *J. Virol.*, 75, 10880-10891.

[95] Rocheleau, J.V. and Petersen, N.O. (2000). Sendai virus binds to a dispersed population of NBD-GD1a. *Biosci. Reports*, 20, 139-155.

[96] Rohde, M., Müller, E., Chhatwal, G.S. and Talay, S.R. (2003). Host cell caveolae act as an entry port group A streptococci. *Cell. Microbiol.*, 5, 323-342.

[97] Sato, Y.; Kaneko, K.; Sasahara, T. and Inoue, M. (2006). Novel pathogenic mechanism in a clinical isolate of *Yersinia enterocolitica* KU14. *J. Microbiol.*, 44, 98-105.

[98] Saukkonen, K.; Burnette, N.W.; Mar, V.L.; Masure, R.H. and Tuomanen, E.I. (1992). Pertussis toxin has eukaryotic-like carbohydrate recognition domains. *Proc. Natl. Acad. Sci. USA,* 89, 118-122.

[99] Schengrund, C-L. (2003). "Multivalent" saccharides: development of new approaches for inhibiting the effects of glycosphingolipid-binding pathogens. *Biochem. Pharmacol.*, 65, 699-707.

[100] Schmidt, M. and Chiorini, J.A. (2006). Gangliosides are essential for bovine adeno-associated virus entry. *J. Virol.,* 80, 5516-5522.

[101] Schmitter, T.; Pils, S.; Weibel, S.; Agerer, F.; Peterson, L.; Buntru, A.; Kopp, K. and Hauck C.R. (2007). Opa proteins of pathogenic *Neisseriae* initiate Src-kinase-dependent or lipid raft-mediated uptake via distinct human CEACAM isoforms. *Infect. Immun.* Online, published 21 May doi:10.1128/IAI.01835-06.

[102] Schneider, B.; Schueller, C.; Utermoehlen, O. and Haas, A. (2007). Lipid microdomain-dependent macropinocytosis determines compartmentation of *Afipia felis. Traffic,* 8, 226-240.

[103] Schraw, W.; Li, Y.; McClain, M.S.; Gisou van der Goot, F. and Cover, T.L. (2002). Association of *Helicobacter pylori* vacuolating toxin (VacA) with lipid rafts. *J. Biol. Chem.*, 37, 34642-34650.

[104] Schwarzmann, G. (2001). Uptake and metabolism of exogenous glycosphingolipids by culture cells. *Cell Develop. Biol.*, 12, 163-171.

[105] Seveau, S.; Bierne, H.; Giroux, S.; Prévost, M-C. and Cossart, P. (2004). Role of lipid rafts in E-cadherin- and HGF-R/Met-mediated entry of *Listeria monocytogenes* into host cells. *J. Cell. Biol.,* 166, 743-753.

[106] Shen, W.; Falahati, R.; Stark, R.; Leitenberg, D. and Ladisch, S. (2005). Modulation of CD4 Th cell differentiation by ganglioside GD_{1a} in vitro, *J. Immunol.,* 175, 4927-4934.

[107] Shimada, Y.; Maruya, M.; Iwashita, S. and Ohno-Iwashita, Y. (2002). The C-terminal domain of perfringolysin O is an essential cholesterol-binding unit targeting to cholesterol-rich microdomains. *Eur. J. Biochem.*, 269, 6195-6203.

[108] Shin, J-S.; Gao, Z. and Abraham, S.N. (2000). Involvement of cellular caveolae in bacterial entry into mast cells. *Science*, 289, 785-787.

[109] Simons, K. and Ehehalt, R. (2002). Cholesterol, lipid rafts, and disease. *J. Clin. Invest.*, 110, 597-603.

[110] Smith, D.C.; Spooner, R.A.; Watson, P.D.; Murray, J.L.; Hodge, T.W.; Amessou, M.; Johannes, L.; Lord, M.J. and Roberts, L.M. (2006). Internalized pseudomonas exotoxin A can exploit multiple pathways to reach the endoplasmic reticulum. *Traffic*, 7, 379-393.

[111] Smith, D.C.; Sillence, D.J.; Falguières, T.; Jarvis, R.M.; Johannes, L.; Lord, M.J.; Platt, F.M. and Roberts, L.M. (2006). The association of Shiga-like toxin with detergent-resistant membranes is modulated by glucosylceramide and is an essential requirement in the endoplasmic reticulum for a cytotoxic effect. *Mol. Biol. Cell*, 17, 1375-1387.

[112] Smith, J.L.; Campos, S.K. and Ozbun, M.A. (2007). Human papillomavirus type 31 uses a caveolin 1- and dynamin 2-mediated entry pathway for infection of human keratinocytes. *J. Virol.*, Online, published 11 July, doi:10.1128/JVI.00988-07.

[113] Srnka, C.A.; Tiemeyer, M.; Gilbert, J.H.; Moreland, M.; Schweingruber, H.; De Lappe, B.W.; James, P.G.; Gant, T.; Willoughby, R.E.; Yolken, R.H.; Nashed, M.A.; Abbas, S.A. and Laine, R.A. (1992). Cell surface ligands for rotavirus: mouse intestinal glycolipids and synthetic carbohydrate analogs. *Virol.,* 190, 794-805.

[114] Stang, E.; Kartenbeck, J. and Parton, R.G. (1997). Major histocompatibility complex class I molecules mediate association of SV40 with caveolae. *Mol. Biol. Cell,* 8, 47-57.

[115] Stuart, A.D.; Eustace, H.E.; McKee, T.A. and Brown, T.D.K. (2002). A novel cell entry pathway for a DAF-using human enterovirus is dependent on lipid rafts. *J. Virol.*, 76, 9307-9322.

[116] Stuart, E.S.; Webley, W.C. and Norkin, L.C. (2003). Lipid rafts, caveolae, caveolin-1, and entry by *Chlamydiae* into host cells. *Exp. Cell Res.*, 287, 67-78.

[117] Sukumaran, S.K.; Quon, M.J. and Prasadarao, N.V. (2002). *Escherichia coli* K1 internalization via caveolae requires caveolin-1 and protein kinase Ca interaction in human brain microvascular endothelial cells. *J. Biol. Chem.*, 277, 50716-50724.

[118] Superti, F.; Hauttecoeur, B.; Morelec, M-J.; Goldoni, P.; Bizzini, B. and Tsiang, H. (1986). Involvement of gangliosides in rabies virus infection. *J. Gen . Virol.*, 67, 47-56.

[119] Tamai, R.; Asai, Y. and Ogawa, T. (2005). Requirement for intercellular adhesion molecule 1 and caveolae in invasion of human oral epithelial cells by *Porphyromonas gingivalis. Infect. Immun.*, 73, 6290-6298.

[120] Thomas, R. and Brooks, T. (2004). Common oligosaccharide moieties inhibit the adherence of typical and atypical respiratory pathogens. *J. Med. Microbiol.*, 53, 833-840.

[121] Thorp, E.B. and Gallagher, T.M. (2004). Requirements for CEACAMs and cholesterol during murine coronavirus cell entry. *J. Virol.,* 78, 2682-2692.

[122] Tong, H.H.; McIver, M.A.; Fisher, L.M. and DeMaria, T.F. (1999). Effect of lacto-N-neotetraose, asialoganglioside-GM1 and neuraminidase on adherence of otitis media-associated serotypes of *Streptococcus pneumoniae* to chinchilla tracheal epithelium. *Microb. Pathog.,* 26, 111-119.

[123] Triantafilou, K. and Triantafilou, M. (2003). Lipid-raft microdomains: key sites for Coxsackievirus A9 infectious cycle. *Virol.,* 317, 128-135.

[124] Triantafilou, K. and Triantafilou, M. (2004). Lipid-raft dependent Coxsackievirus B4 internalization and rapid targeting to the Golgi. *Virol.*, 326, 6-19.

[125] Triantafilou, M.; Manukyan, M.; Mackie, A.; Morath, S.; Hartung, T.; Heine, H. and Triantafilou, K. (2004). Lipoteichoic acid and toll-like receptor 2 internalization and targeting to the golgi are lipid raft-dependent. *J. Biol. Chemistry,* 279, 40882-40889.

[126] Tsai, B.; Gilbert, J.M.; Stehle, T.; Lencer, W.; Benjamin, T.L. and Rapoport, T.A. (2003). Gangliosides are receptors for murine polyoma virus and SV40. *EMBO J.*, 22, 4346-4355.

[127] Uotsu, N.; Nishikawa, A.; Watanabe, T.; Ohyama, T.; Tonozuka, T.; Sakano, Y. and Oguma, K. (2006). Cell internalization and traffic pathway of *Clostridium botulinum* type C neurotoxin in HT-29 cells. *Biochim. Biophys. Acta*, 1763, 120-128.

[128] van Alphen, L.; Geelen-van den Broek, L.; Blaas, L.; van Ham, M. and Dankert, J. (1991). Blocking of fimbria-mediated adherence of *Haemophilus influenzae* by sialyl gangliosides. *Infect. Immun.,* 59, 4473-4477.

[129] Waddell, T.; Cohen, A. and Lingwood, C.A. (1990). Induction of verotoxin sensitivity in receptor-deficient cell lines using the receptor glycolipid globotriosylceramide. *Proc. Natl. Acad. Sci. USA*, 87, 7898-7901.

[130] Waheed, A.A.; Shimada, Y.; Heijnen, H.F.G.; Nakamura, M.; Inomata, M.; Hayashi, M.; Iwashita, S.; Slot, J.W. and Ohno-Iwashita, Y. (2001). Selective binding of perfringolysin O derivative to cholesterol-rich membrane microdomains (rafts). *Proc. Natl. Acad. Sci. USA,* 98, 4926-4931.

[131] Walch, M.; Eppler, E.; Dumrese, C.; Barman, H.; Groscurth, P. and Ziegler, U. (2005). Uptake of granulysin via lipid rafts leads to lysis of intracellular *Listeria innocua*. *J. Immunol.*, 174, 4220-4227.

[132] Watarai, M.; Makino, S-i.; Fujii, Y.; Okamoto. K. and Shirahata, T. (2002). Modulation of *Brusella*-induced macropinocytosis by lipid rafts mediates intracellular replication. *Cell. Microbiol.,* 4, 341-355.

[133] Werling, D.; Hope, P.C.; Collins, R.A.; Taylor, G. and Howard, C.J. (1999). Involvement of caveolae in the uptake of respiratory syncytial virus antigen by dendritic cells. *J. Leukocyte Biol.*, 66, 50-58.

[134] Wielgosz, M.M.; Rauch, D.A.; Jones, K.S.; Ruscetti, F.W. and Ratner, L. (2005). Cholesterol dependence of HTLV-I infection. *AIDS Research and Human Retroviruses*, 21, 43-50.

[135] Wooldridge, K.G.; Williams, P.H. and Ketley, J.M. (1996). Host signal transduction and endocytosis of *Campylobacter jejuni*. *Microbial Pathog.,* 21, 299-305.

[136] Zhuang, M.; Oltean, D.I.; Gomez, I.; Pullikuth, A.K.; Soberon, M.; Bravo, A. and Gill, S.S. (2002). *Heliothis virescens* and *Manduca sexta* lipid rafts are involved in Cry1A toxin binding to the midgut epithelium and subsequent pore formation. *J. Biol. Chem.*, 277, 13863-13872.

In: Glycolipids: New Research
Editor: Daiki Sasaki, pp. 53-78

ISBN: 978-1-60456-216-3

Chapter II

The Importance of Gangliosides in Mast Cell Biology

Adriana Maria Mariano Silveira e Souza, Maria Célia Jamur and Constance Oliver*
Department of Cell and Molecular Biology and Pathogenic Bioagents
Faculdade de Medicina de Ribeirão Preto – University of São Paulo
Ribeirão Preto, SP – Brazil
Reviewed by Maria Cristina Roque-Barreira
Department of Cell and Molecular Biology and Pathogenic Bioagents,
School of Medicine of Ribeirão Preto,
University of São Paulo, Ribeirão Preto, SP, Brazil

Abstract

Gangliosides are complex glycosphingolipids present in all eukaryotic cells. They are composed of a hydrophobic ceramide backbone anchored in the lipid bilayer of plasma membrane and a polar head group with hydrophobic oligosaccharide chains and one or more residues of *N*-acetylneuraminic acid (sialic acid) at the termini that protrude into the extracellular space. Ganglioside species vary one from another in their ceramide composition and/or in their carbohydrate content. The kinds and amounts of gangliosides expressed in the plasma membrane are cell and tissue type dependent. Ganglioside composition may change during embryonic development, tissue differentiation, aging and degeneration or malignant transformation. Gangliosides have been recognized to play important roles in normal physiological functions as well as in pathological conditions, including tumor progression. They modulate many biologic processes, such as differentiation, proliferation, adhesion, migration, immune response, and act as receptors for viruses and bacterial toxins. Gangliosides also participate in apoptosis and signal transduction in several cell types. In mast cells, gangliosides may function in

* Address correspondence to: Dr. Constance Oliver, Department of Cell and Molecular Biology and Pathogenic Bioagents, School of Medicine of Ribeirão Preto, University of São Paulo, Av. Bandeirantes 3900. 14049-900 - Ribeirão Preto, SP. Brazil. Phone: 55-16-3602-3217. FAX: 55-16-3633-1786. e-mail: coliver@fmrp.usp.br

modulating secretory events as well as participating in mast cell development and recruitment. Mast cells have long been known to play a critical role in inflammatory and allergic reactions and contribute to host resistance. Following mast cell activation by cross-linking of the high affinity IgE surface receptor (FcεRI) on the cell surface a number of inflammatory mediators are released. Previous studies using the monoclonal antibody (mAb) AA4, which recognizes two α-galactosyl derivatives of ganglioside GD_{1b} that are specific to rodent mast cells showed that these gangliosides are located in proximity to FcεRI. Binding of mAb AA4 to the gangliosides on the surface of mast cells produces morphological and biochemical changes similar to those seen with activation of FcεRI, but without histamine release. The expression of these gangliosides on the mast cell surface has also been shown to be related to mast cell maturation. Committed mast cell precursors do not express the gangliosides recognized by mAb AA4, but the gangliosides begin to be expressed on the cell surface jointly with FcεRI in very immature mast cells and continue to be expressed by mast cells in all stages of maturation. This chapter will present a short review on the biological functions of gangliosides and discuss some important evidence on the role of gangliosides in modulating FcεRI and in the maintenance of mast cell structure and function.

Introduction

The Singer-Nicolson fluid mosaic model for cell membranes (1972), the most accepted model of biological membranes, says that the basic structure of cell membranes is a double layer of lipids. Cell membranes delimit cells and organelle, act as barrier to the passage of polar molecules and ions and contain membrane proteins (Brown and London, 1998; Simons and Toomre, 2000; Harder, 2001). Membrane lipids are amphipathic, i. e., one end of the molecule is hydrophobic and the other hydrophilic. The hydrophobic interactions of lipids with each other and their hydrophilic interactions with water direct their packing into the sheets termed the lipid bilayer. This structure is formed mainly from three different classes of lipids: *glycerophospholipids*, in which the hydrophobic regions are composed of two fatty acids joined to glycerol; *sterols*, compounds characterized by a rigid system of four fused hydrocarbon rings; and *sphingolipids*, which in their simplest form are composed of one molecule of the long-chain amino alcohol sphingosine linked through an amide group, to a long-chain fatty acid. The resulting compound, a ceramide, is the structural parent of all sphingolipids. Ceramides usually include a polar head group esterified to the terminal hydroxyl of the sphingosine. The head groups define the sphingolipid classes, with a phosphorylcholine or a phosphoroethanolamine group found in sphingomyelin and carbohydrates in glycosphingolipids (Nelson and Cox, 2000; Pomorski *et al.*, 2001; Reynolds *et al.*, 2004).

Glycosphingolipids or, simply glycolipids, are a large and heterogeneous family of the sphingolipid constituent of cell membranes that are enriched in the plasma membrane. They are embedded in the extracytoplasmic leaflet of these membranes through their ceramide moiety and expose the carbohydrate-containing hydrophobic portion to the extracytoplasmic side. Glycosphingolipids show heterogeneity not only in type, number, and linkage of sugar residues in the oligosaccharide chain but also in their ceramide moieties. The structure of ceramide, especially the fatty acid moieties, can influence the localization and function of glycosphingolipids in the plasma membrane. Animal cells and several plant cells contain at least one glycosphingolipid class like cerebrosides, sulfatides, globosides and gangliosides in their membranes (Hakomori, 1993; Gillard *et al.*, 1993; Maccioni *et al.*, 1999; Knoll *et al.*, 2000; Osterbye *et al.*, 2001; King and Marchesini, 2002).

Cerebrosides have a single carbohydrate group linked to ceramide. The most common carbohydrate is galactose (galactocerebrosides), with glucose (glucocerebrosides) being linked less frequently. Galactocerebrosides are found predominantly in neuronal cell membranes. By contrast, glucocerebrosides are found in the plasma membranes of cells in non neuronal tissues. They also represent intermediates in the synthesis or degradation of more complex glycosphingolipids. Excess accumulation of glucocerebrosides due to a deficiency jn glucocerebrosidase is observed in Gaucher's disease. The sulfuric acid esters of galactocerebrosides are the sulfatides. They are synthesized primarily by oligodendrocytes in the central nervous system. Excess accumulation of sulfatides due to a lack of the lysosomal enzyme sulfatide sulfatase (arylsulfatase A) is observed in sulfatide lipidosis (metachromatic leukodystrophy). Globosides represent cerebrosides that contain additional carbohydrates, usually galactose, glucose or *N*-acetyl-galactosamine. Systemic accumulation of globosides, particularly globotriaosylceramide in lysosomes of the vascular endothelial cells of most organs due to the deficient activity of alpha-galactosidase A is observed in Fabry's disease.

Gangliosides are the group of glycosphingolipids that show the greatest structural variation and also the most complex structure. The name gangliosides was first applied by the german scientist Ernst Klenk in 1942 to lipids newly isolate from ganglion cells of the brain. These lipids can amount to 6% of the weight of lipids from brain, but they are found at low levels in all animal tissues, primarily as plasma-membrane components. Gangliosides have oligosaccharides as their polar head groups and one or more residues of *N*-acetylneuraminic acid (Neu5Ac), also called sialic acid, at their termini. Sialic acid gives gangliosides the negative charge at pH 7 that distinguishes them from other glycosphingolipids. They are classified into several subclasses depending on the structure of the oligosaccharide backbones (hematoside-series, ganglio-series, lacto and neolacto-series, and globo and isoglobo-series; Table I). The nomenclature system suggested by Svennerholm, in 1980, for the ganglio-series gangliosides gives an idea about the gangliosides structure: the letter G refers to "ganglioside", M, D, T and Q indicate that the molecule contains mono, di, tri or tetra sialic acid residues. The number subscripts assigned to the individual compound, that initially referred to its migration order in a certain chromatographic system (Svennerholm, 1963), refers to the carbohydrate sequence that is attached to ceramide; 1 stands for Gal-GalNAc-Gal-Glc-ceramide[1], 2 for GalNAc-Gal-Glc-ceramide[2] and 3 for Gal-Glc-ceramide[3]. Despite the subsequent introduction of a more systematized official nomenclature for these molecules, the Svennerholm nomenclature system has persisted, not only because of its historical value, but also because investigators prefer to call specific molecules by a common name that is easy to remember. A list of these abbreviations is given in Table II (IUPAC-IUB, 1998; Radin, 1999; Varki *et al.,* 1999; Nelson and Cox, 2000; Hideharu and Makoto, 2001; King and Marchesini, 2002; Christie, 2003).

[1] Galactose-*N*-acetyl-galactosamine-galactose-glucose-ceramide
[2] *N*-acetyl-galactosamine-galactose-glucose-ceramide
[3] Galactose-glucose-ceramide

Table I. Fundamental carbohydrate structure of gangliosides (Adapted from Hideharu and Makoto, 2001)

- hematoside-series
  - Galβ(1→1)Cer
  - Galβ(1→4)Glcβ(1→1)Cer

- ganglio-series
  - GalNAcβ(1→4)Galβ(1→4)Glcβ(1→1)Cer
  - Galβ(1→3)GalNAcβ(1→4)Galβ(1-4)Glcβ(1→1)Cer

- lacto and neolacto-series
  - [Galβ(1→3)GlcNAcβ(1→3)]nGalβ(1→4)Glcβ(1→1)Cer
  - [Galβ(1→4)GlcNAcβ(1→3)]nGalβ(1→4)Glcβ(1→1)Cer

- globo and isoglobo-series
  - GalNAcβ(1→3)Galα(1→4)Galβ(1→4)Glcβ(1→1)Cer
  - GalNAcβ(1→3)Galα(1→3)Galβ(1→4)Glcβ(1→1)Cer

Table II. Some abbreviations using the Svennerholm system (Adapted from IUPAC-IUB - Nomenclature of glycolipids. Recommendations 1997)

Structure	Abbreviation
Neu5Acα3Galβ4GlcCer	GM_3
GalNAcβ4(Neu5Acα3)Galβ4GlcCer	GM_2
Galβ3GalNAcβ4(Neu5Acα3)Galβ4GlcCer	GM_{1a}
Neu5Acα3Galβ3GalNAcβ4Galβ4GlcCer	GM_{1b}
Neu5Acα8Neu5Acα3Galβ4GlcCer	GD_3
GalNAcβ4(Neu5Acα8Neu5Acα3)Galβ4GlcCer	GD_2
Neu5Acα3Galβ3GalNAcβ4(Neu5Acα3)Galβ4GlcCer	GD_{1a}
Galβ3GalNAcβ4(Neu5Acα8Neu5Acα3)Galβ4GlcCer	GD_{1b}

Based on the structure of gangliosides, the chemistry of their synthesis can be divided into two categories. One is the mode common to all kinds of gangliosides, the introduction of sialic acid and sphingolipid, and the other, the construction of the individual oligosaccharide backbones is specific to each subclass of ganglioside (Hideharu and Makoto, 2001). The biosynthesis of gangliosides involves sequential actives of glucosyltransferases and sialyltransferases, and begins with the ceramide, which is synthesized on the membranes of endoplasmic reticulum by the enzyme serine palmitoyl transferase. After transport to the Golgi apparatus, glucosylceramide synthase transfers a glucose residue to ceramide to produce glucosylceramide. This reaction occurs on the cytosolic surface of the Golgi. Glucosylceramide is then transferred to the lumenal leaflet of the Golgi, where it is modified by the addition of a galactose moiety to produce lactosylceramide, the precursor to most gangliosides. The attachment of a sialic acid produces the simple ganglioside GM_3, which serves as an essential core structure during biosynthesis of the more complex oligosaccharide chains found on gangliosides. Finally, the gangliosides are transferred to the plasma membrane by a transport system involving vesicle formation and/or ganglioside transfer proteins (Gammon and Leeden, 1985; Kolter *et al.*, 2002; Allende and Proia, 2002).

Gangliosides, as well as other membrane constituents, are removed from the plasma membrane by endocytosis and are recycled to plasma membrane or transported to lysosomes for degradation. Important sorting events occur at the stage of early endosomes. Recycling membrane components are concentrated in the tubular extensions of early endosomes and subsequently included in recycling vesicles, while soluble components, such as dissociated ligands, remain in the vesicular portion of the endosome and are transported to lysosomes (Möbius *et al.*, 1999). The gangliosides are delivered in the form of small vesicles to lysosomes for degradation. Glycosphingolipids are degraded in a stepwise manner by specific acid hydrolases. Some of these enzymes need the assistance of small nonenzymatic cofactors called sphingolipid activator proteins (SAPs) for the degradation of glycosphingolipids with short oligosaccharide head groups. Factors such as membrane curvature, membrane pressure and lipid composition enable the efficient degradation of glycosphingolipids in the small lysosomal vesicles (Werth *et al.*, 2001; Allende and Proia, 2002).

Genetic defects, such as deficiencies in lysosomal enzymes, have been described for many of the steps of gangliosides and glycosphingolipids degradation, resulting in a family of disorders related to sphingolipid storage (Tifft and Proia, 2000). In Tay-Sachs disease there is an excess accumulation of ganglioside GM_2 caused by insufficient activity of β-hexosaminidase A. Accumulation of GM_1 ganglioside due to deficiency of the lysosomal hydrolase, acid β-galactosidase, is observed in GM_1 gangliosidosis (King and Marchesini, 2002). Table III lists some examples of sphingolipid storage diseases, its causes and principal symptoms.

These disorders generally involve severe neurodegeneration owing, in part, to the relatively high rate of gangliosiode synthesis in the brain. New strategies of treatment are being investigated that use small-molecule inhibitors of glucosylceramide synthase, the key enzyme for glycosphingolipid synthesis, to block endogenous cellular ganglioside production and lower the synthesis rates of glycosphingolipids and, thus, reduce the amount of lysosomal glycosphingolipid accumulation in these storage diseases (Platt *et al.*, 1997; Liu *et al.*, 1999; Tifft and Proia, 2000; Abe *et al.*, 2000; Birklé *et al.*, 2003).

Gangliosides have been shown to have important regulatory roles in the normal physiological process, as well as in pathological conditions, including tumor onset and progression (Hakomori, 1996; Yamashita *et al.*, 1999). The composition of ganglioside species expressed in cells is cell type, tissue and species-specific as well as specifically changed during development and differentiation, ageing, degeneration and malignant transformation (Hakomori, 1981; Svennerholm *et al.*, 1989; Svennerholm *et al.*, 1991; Yu, 1994; Metelmann *et al.*, 2001). Recently, cumulative evidence indicates that tumor-associated gangliosides are a result of initial oncogenic transformation and play a key role in the induction of invasion and metastasis. The concept of ganglioside-dependent promotion of tumor progression has been developed in conjunction with clinico-pathological studies that have demonstrated that there are some gangliosides species with relatively simple structures that show very restricted expression in normal tissues and markedly enhanced expression in particular malignant tumor. Certain gangliosides, such as GD_3 or GM_2, promote tumor-associated angiogenesis; others strongly regulate cell adhesion/motility and thus initiate tumor metastasis (Birklé *et al.,* 2003).

Table III. Disorders associated with abnormal shingolipids metabolism (Adapted from King and Marchesini, 2002)

Disorder	Enzyme deficiency	Accumulating substance	Symptons
Tay Sachs disease	Hexosaminidase A	GM_2 ganglioside	mental retardation, blindness, early mortality
Gaucher's disease	Glucocerebrosidase	Glucocerebroside	hepatosplenomegaly, mental retardation in infantile form, long bone degeneration
Fabry's disease	α-Galactosidase A	Globotriaosylceramide	kidney failure, skin rashes
GM_1 gangliosidosis	β-galactosidase	GM_1 ganglioside	mental retardation, skeletal abnormalities, hepatomegaly

The biological importance of gangliosides in cellular regulation is well recognized (Fishman, 1986; Hannun and Bell, 1989; Igarashi *et al.,* 1989; Hakomori, 1990, Oliver *et al.,* 1992; Palestini *el al.,* 2000; Metelmann *et al.,* 2001; Allende and Proia, 2002). Over the years, gangliosides have been implicated in a number of cell surface mediated events, such as cell-cell recognition processes and cell-substratum interaction, as well as acting as modulators of transmembrane signaling. They can also modulate cell growth, differentiation, proliferation, migration and adhesion, immune response and apoptosis. Gangliosides can act as receptors for viruses and bacterial toxins and also modulate receptor functions in several cell types, including mast cells (Oliver *et al.,* 1992; Palestini *et al.,* 2000; Allende and Proia, 2002; Hakomori 2002). Some examples on the function of gangliosides as receptors or in modulating receptor functions are given below:

a) Cholera toxin (CT) consists of noncovalently associated A and B subunits. Five identical polypeptides assemble into a pentameric ring to form the B-subunit that binds to a membrane ganglioside GM_1 at the plasma membrane of enterocytes. Upon binding, the A-subunit is cleaved into A1 and A2 chains. The A2 chain protrudes with its C terminal through the central pore in the B-ring and has a KDEL motif facing the membrane. The A1 chain enters the cytosol, where it activates the G protein $G_s\alpha$ through an ADP-ribosylation reaction that acts to lock the G protein in its GTP-bound form, thereby continually stimulating adenylate cyclase to produce cAMP. The high cAMP levels stimulate enterocytes to pump large amount of Cl^- into the intestinal contents. Water, Na^+ and other electrolytes follow due to the osmotic and electrical gradients caused by the loss of Cl^-, leading to diarrhea, loss of electrolytes, and dehydration that are characteristic of cholera. (Sixma *et al.,* 1993; Zhang *et al.*, 1995; Fujinaga *et al.,* 2003);

b) The EGF receptor is a transmembrane glycoprotein that regulates cell growth in both normal and cancerous cells. The binding of EGF to its receptor results in autophosphorylation of tyrosine residues on the receptor and in activation of the Ras, Raf and the mitogen-actived protein kinase (MAPK) cascade that ends in cell proliferation. Gangliosides are involved in modulating the EGF-mediated proliferation pathways (Rebbaa *et al.,* 1996; Li *et al.,* 2000; Meuillet *et al.,* 2000; Allende and Proia, 2002). GM_3 ganglioside has been shown to inhibit EGF-mediated signaling by directly interacting with the extracellular domain of the receptor. This interaction specifically inhibits EGF receptor autophosphorylation and dimerization, and reduces the activities of MAPK and phosphatidylinositol-3-OH kinase. The interaction of GM_3 and the EGF receptor requires receptor glycosylation (Rebbaa *et al.,* 1996; Wang *et al.,* 2001; Miljan *et al.,* 2002);

c) The insulin receptor mediates signaling pathways that are essential for glucose homeostasis and cell proliferation. Its activity is also inhibited by GM_3 ganglioside. Direct addition of GM_3 to cultured adipocytes suppresses phosphorylation of the receptor, the downstream substrate IRS-1 (insulin receptor substrate-1) and, as a consequence, reduces the uptake of glucose (Tagami *et al.,* 2002).

Another unique and important feature of gangliosides is their enrichment along with sphingomyelin, cholesterol, and a series of functionally related proteins within plasma membrane domain, also called lipid rafts.

Gangliosides Are Lipid Rafts Constituents

Developments in cell biology amended the original notion of cell membranes, inferred from the fluid-mosaic model (Singer and Nicolson, 1972). The model predicted free rotational and lateral diffusion of proteins and lipids within the plane of the membrane resulting in their random distribution. A number of lines of evidence indicating a non-random distribution of complex lipids and proteins in cell membranes were presented in seventies and eighties. However, it was necessary to wait until the improvement of membrane fractionation techniques that occurred in the 1990`s, in order to have the initial information on membrane micro-domains (Thompson *et al.,* 1985; Smart *et al.,* 1995; Rodgers and Rose, 1996; Iwabuchi *et al.,* 1998; Dolo *et al.,* 2000; Dráber and Dráberová, 2001).

Lipid rafts (Figure 1) are dynamic subdomains of the plasma membrane and other intracellular membranes, such as Golgi membranes, that contain high levels of cholesterol, sphingolipids such as sphingomyelin and gangliosides.

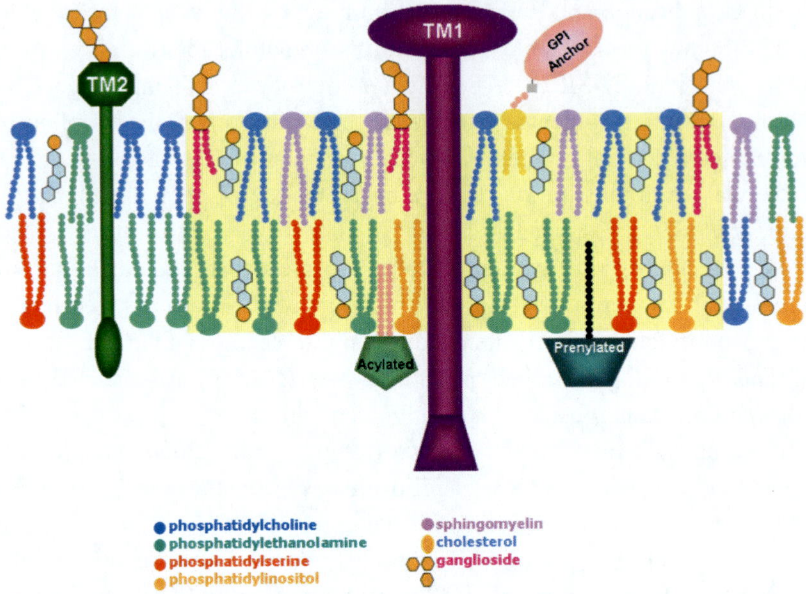

Figure 1. Graphic representation of lipid raft model. Lipid raft, background in yellow, is enriched cholesterol, sphingolipids and gangliosides. In the outer leaflet there are GPI anchored proteins, gangliosides and sphingomyclin. Thc cytoplasmic leaflet contains prenylated proteins, proteins carrying saturated fatty acyl chains, phosphatidylserine and phosphatidylethanolamine. Cholesterol is present in both leaflets and fills the space under the head groups of sphingolipids. TM1, TM2: transmembrane proteins 1 and 2.

Lipid rafts selectively concentrate glycosylphosphatidylinositol (GPI)-anchored proteins on their outer side and proteins anchored by satured palmitoyl or myristoyl groups and cholesterol-binding proteins on the cytoplasmic side (Simons and Ikonen, 1997, Melkonian et al., 1999; Pike, 2003; Rajendran and Simons, 2005). Because of their lipid composition, with a preponderance of saturated hydrocarbon chains that allows cholesterol to be tightly intercalated, lipid rafts are highly organized and exist in a liquid-ordered (l_o) phase, different from the rest of the plasma membrane which consists mainly of phospholipids in a liquid-disordered (l_d) phase. Membrane microdomains containing tightly packed long saturated acyl chains of glycosphingolipids and cholesterol are characterized by high melting temperature and resistance to solubilization in non-ionic detergents such as Triton X-100, at low temperature. These properties were first demonstrated by Brown and Rose, 1992, while investigating differential membrane trafficking in polarized epithelial cells. Because of these properties, lipid rafts are also known as detergent-resistant membranes (DREMs) or glycosphingolipids-enriched membranes (GEMs) (Brown and London, 1998; Dráber and Dráberová, 2001; Pike, 2004).

These membrane microdomains have been implicated in diverse cellular processes, including endocytic events, such as viral entry, membrane trafficking, signal transduction, and cell adhesion, although the most important role of rafts at the cell surface may be their function in cell signaling by virtue of their affinity for signaling proteins, including transmembrane receptors, proteins anchored by GPI, G proteins, RhoA and Src kinases (Allende and Proia, 2002). They are also thought to be important sites for protein-tyrosine-kinase-mediated protein-protein interactions that are involved in the initiation of receptor signaling pathways (Simons and Toomre, 2000; Pike, 2004; Gilfillan and Tkaczyk, 2006). It is well known that, in the case of tyrosine kinase receptors, adaptors, scaffolding proteins and enzymes are recruited to the cytoplasmic side of the plasma membrane as a result of ligand binding (Hunter, 2000). Lipid rafts are also considered to be platforms for receptors activated by ligand binding. If receptor activation takes place in an ordered lipid raft, the signaling complex is protected from other proteins, such as membrane phosphatases, localized in the disordered region of the plasma membrane, that otherwise could affect the signaling process (Simons and Toomre, 2000; Holowka *et al.,* 2005). Lipid rafts are implicated in the function of diverse signaling pathways such as those mediated by growth factors, morphogens, integrins (Simons and Toomre, 2000) and antigen receptors on immune cells (Field *et al.,* 1995; Xavier *et al.,* 1998; Miceli *et al.,* 2001; Cheng *et al.,* 2001; Holowka and Baird, 2001).

Lipid rafts are dynamic in that both proteins and lipids can move in and out of raft domains with different partitioning kinetics (Rajendran and Simons, 2005). Despite much evidence supporting the existence of raft domains, the raft concept is still being debated (Edidin, 2003). The controversy is largely due to the lack of standardized methodology for lipid raft studies and the difficulty in proving definitively that rafts exist in living cells, since that lipid rafts are too small to be resolved by standard light microscopy. However, recent advances in imaging and studies with improved methodology are now providing insights into the existence and behavior of lipid rafts (Varma and Mayor, 1998; Simons and Toomre, 2000; Wilson *et al.,* 2000; Pralle *et al.,* 2000, Parton and Richards, 2003).

Mast Cells

Mast cells, like blood cells, are derived from bone marrow pluripotent hematopoietic stem cells but, unlike blood cells, they leave the bone marrow as progenitors and migrate into virtually all vascularized tissues to complete their differentiation under the influence of factors present at each tissue site. It is the microenvironment surrounding the mast cells that determines their mature phenotype (Kitamura, 1989; Metcalfe *et al.,* 1997; Galli, 2000; Jamur *et al.,* 2001; Galli *et al.*, 2005; Jamur *et al.,* 2005; Gilfillan and Tkaczyk, 2006). These cells normally reside close to epithelia, associated with blood vessels, nerves, and found within mucosal surfaces of the gastrointestinal and respiratory tracts, in the skin, near muscle cells and mucus-producing glands. In some species, including murine rodents, mast cells also occur within mesothelium-lined cavities, such as the peritoneal cavity (Robbie-Ryan and Brown, 2002; Galli *et al,* 2005). Although mast cells share many characteristics, it has been known since their discovery that they do not represent a homogeneous population. Based on several differences such as anatomic location, morphology, histochemistry and function, rodent and human mast cells are classified as connective tissue mast cells or mucosal mast cells (Metcalfe *et al.,* 1997; Yong, 1997).

Mast cells were first recognized by Paul Ehrlich in 1876. His microscopic observations revealed that mast cells had a high content of cytoplasmic granules, which stain metachromatically with basic aniline dyes such as toulidine blue. These granules were later shown to contain potent biologically active mediators which are released following cell activation (Yong, 1997). Because of this mediator release, mast cells have long been recognized to play a direct and critical role in allergic and inflammatory reactions (Siraganian, 1988, Williams and Galli, 2000). In allergic diseases, mast cells exert both local and systemic responses, including allergic rhinitis and anaphylaxis, as well as on late phase response and chronic asthma (Robbie-Ryan and Brown, 2002). Mast cell mediators are also related to many chronic inflammatory conditions, including stress-induced intestinal ulceration, rheumatoid arthritis and Crohn's disease. Besides the roles in pathological conditions, the biological functions of mast cells include roles in innate immunity, involvement in host defense mechanisms against parasites, immunomodulation of the immune system, tissue repair and angiogenesis (Metcalfe *et al.,* 1997; Abraham and Malaviya, 1997). Recently, mast cells have gained new importance as immunoregulatory cells with the discovery that they are a major source of cytokines and chemokines (Galli *et al.,* 2005).

Although mast cells may be activated by a number of stimuli and pathways, the major mechanism for their activation is through the high-affinity receptor for immunoglobulin E (FcεRI), present in the plasma membrane of mast cells as well as basophils. Mast cell activation is initiated by the binding of oligomeric antigens to receptor-bound IgE, which crosslinks FcεRI and results in its aggregation. The activation of FcεRI results in a complex series of biochemical and morphological events, that culminate in the release of three classes of mediators: *preformed mediators*, which are stored in secretory granules, such as histamine, heparin, β-hexosaminidase, neutral proteases, acid hydrolases, major basic protein, carboxypeptidases, and some cytokines and growth factors; *newly formed lipid mediators,* such as prostaglandins and leukotrienes; and *newly synthesized mediators*, that include growth factors, cytokines and chemokines (Metzger, 1992; Siraganian, 2003; Oliver *et al.,* 2007).

FcεRI is expressed on the surface of mast cells (and basophils) as a heterotetrameric structure composed of one α subunit with an extracellular domain that binds IgE, a four transmembrane-spanning β subunit, and two identical disulphide linked γ subunits (Metzger, 1992; Kinet, 1999; Simons and Toomre, 2000; Rivera *et al.,* 2002). The β subunit serves as an important amplifier of IgE and antigen induced signaling events. Furthermore, the γ subunits are essential for initiating signaling events downstream of FcεRI (Nadler *et al.,* 2000; Rivera *et al.,* 2002). The carboxyl terminal cytoplasmic domains of both β and γ subunits contain an immunoreceptor tyrosine-based activation motif (ITAM), common to all multisubunit immune recognition receptors, that is critical for cell activation. Because the receptor subunits lack any known enzymatic activity, FcεRI must rely on associated molecules for transducing intracellular signals (Simons and Toomre, 2000; Rivera *et al.,* 2001a; Kimura *et al.,* 2001).

The first recognized biochemical event of the cytoplasmic signal transduction cascade, after crosslinking of two or more FcεRI, involves tyrosine phosphorylation usually by Lyn, a member of the Src family of tyrosine kinases, of two conserved tyrosine residues within the ITAMs of both β and γ subunits of the receptor. The tyrosine-phosphorylated ITAMs create a novel binding surface that is recognized by additional cytoplasmic signaling molecules, such as the protein tyrosine kinase Syk, a member of Syk/ZAP-70 family, which binds mainly to the γ subunit, via its tandem Src homology 2 (SH2) domains. This interaction results in a conformational change of Syk, followed by its activation and autophosphorylation, that results in an increased kinase activity that rapidly shifts the equilibrium of the cell from a resting state (where phosphorylation and dephosphorylation activies are approximately equal) to an activated state (where phosphorylation activity increases exponentially and cannot be counteracted by dephosphorylation). This Syk mediated signal amplification results in a direct or indirect activation of several proteins, including linker for activation of T cells (LAT), Vav, phospholipase C-γ1 (PLC-γ1) and PLC-γ2. Finally, downstream activation results in an increase of intracellular calcium levels, activation of other enzymes and adaptors, rearrangement of the cytoskeleton and immediate release of preformed and newly formed inflammatory mediators, as well as other cellular responses including sythesis and release of newly synthesized mediators (Kinet, 1999; Rivera *et al.,* 2002; Siraganian, 2003).

Gangliosides and Mast Cells

Despite numerous studies on mast cell activation through FcεRI, the detailed mechanism by which cross-linking promotes this initial phosphorylation by Lyn and how and/or where these proteins interact remains poorly understood (Pyenta *et al.,* 2001; Dráber and Dráberová, 2001; Young *et al.,* 2003).

IgE signaling was initially thought to be based solely on protein-protein interactions. In this model, receptor phosphorylation occurs as a consequence of Lyn that is constitutively associated with β subunit of FcεRI. This preassociated Lyn is hypothesized to mediate the tyrosine phosphorylation of β and γ chains of the receptors by a process of transphosphorylation when the receptors are brought into proximity upon aggregation (Pribluda *et al.,* 1994; Metzger, 1999). An aggregation-dependent-model of FcεRI phosporylation proposes that after cross-linking, FcεRI moves into specialized membrane rafts and that receptor phosporylation by Lyn occurs within these lipid domains. Thus plasma membrane structure may play a critical role in facilitating the coupling between Lyn and FcεRI, during mast cell signal transduction (Baird *et al.,* 1999; Rivera *et al.,* 2001a). Much of the early evidence about the role of lipid rafts in signal transduction came from IgE signaling during the allergic immune response (Field *et al.,* 1995; Sheets *et al.,* 1999). The first hint came from the finding that FcεRI is soluble in Triton X-100 at steady state, but becomes insoluble in low concentrations of this detergent after crosslinking (Field *et al.,* 1995). Moreover, in unstimulated cells, FcεRI is loosely dispersed throughout the plasma membrane, but upon activation rapidly aggregates (Stump *et al.,* 1989; Wilson *et al.,* 2001) and can then be found, when examined by fluorescence microscopy, on the cell surface in association with GM_1 ganglioside (Wilson *et al.,* 2004) and GPI-anchored proteins, molecules typically associated with lipid rafts (Stauffer and Meyer, 1997; Holowka *et al.,* 2000).

Gangliosides seem to be structurally important for lipid raft assembly and function. The rigid structural nature of the ceramide anchor in gangliosides, coupled with the ability of sphingolipids to associate with cholesterol, is thought to drive the assembly of lipid rafts (Simons and Toomre, 2000; Kolter *et al.,* 2002). In rodent mast cells, gangliosides, more precisely the α-galactosyl derivatives of the gangliosides GD_{1b}, may participate in mast cell development and recruitment as well as modulating secretory events (Oliver *et al.,* 1992; Minoguchi *et al.,* 1994; Swaim *et al.,* 1994; Stephan *et al.,* 1997; Jamur *et al.,* 1997; Jamur *et al.,* 2001; Jamur *et al.,* 2005). Alpha-galactosyl derivatives of the gangliosides GD_{1b} (Figure 2), which were originally isolated from RBL-2H3 cells (Guo *et al.*, 1989), a cell line with homology to mucosal mast cells, are unique gangliosides present on the surface of rodent mast cells (Guo *et al.,* 1989; Oliver *et al.,* 1992) that are specifically recognized by mAb AA4 (Guo *et al.*, 1989). The mAb AA4 was originally identified as one of a several antibodies that were produced by immunizing mice with RBL-2H3 cells and then selected due to their ability to inhibit histamine release from RBL-2H3 cells in a dose and time dependent manner (Basciano *et al.,* 1986). These gangliosides are evenly distributed on the surface of RBL-2H3 cells (Figure 3). When RBL-2H3 cells are incubated in the presence of mAb AA4, the gangliosides cap on the cell surface. Little internalization of the gangliosides is seen (Figure 4).

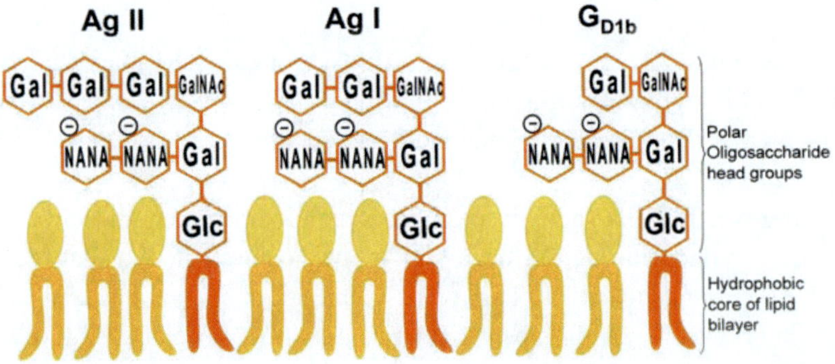

Figure 2. Graphic representation of ganglioside GD_{1b} and α-galactosyl derivatives of ganglioside GD_{1b} (antigen I and antigen II). MAb AA4 recognizes both antigens I and II that are derived from ganglioside GD_{1b}.

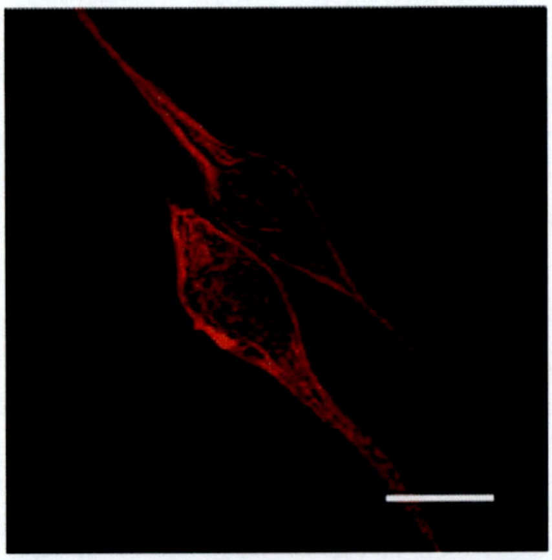

Figure 3. The RBL-2H3 cells highly express the derivatives of the GD_{1b} ganglioside on their surface. The cells were immunolabeled with mAb AA4 and the antibody dectected with donkey anti-mouse IgG coupled to Alexa 594. Samples were examined using a Leica TCS-NT laser scanning confocal microscope. Bar equals 20 μm.

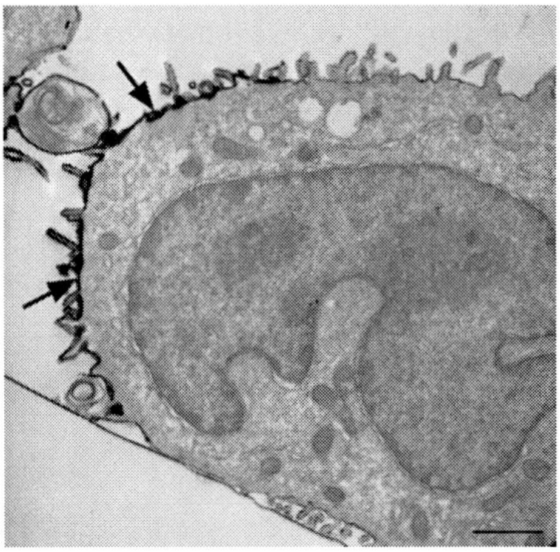

Figure 4. Incubation of RBL-2H3 cells with mAb AA4 induces capping (arrows) of the gangliosides on the cell surface. RBL-2H3 cells were incubated with mAb AA4 conjugated to horseradish peroxidase and after fixation the cells were incubated in reaction medium containing H_2O_2 and diaminobenzidine. The cells were then processed for electron microscopy. Bar equals 1 μm.

Table IV. mAb AA4 Recognizes Only Mast Cells (adapted from Oliver *et al.,* 1992)

Tissue section	Cell stained by mAb AA4
Adrenal gland	Mast Cells
Blood	No Cells Stained
Bone marrow	Mature Mast Cells Immature Mast Cells)
Brain	Mast Cells
Duodenum	Mast Cells
Epididymis	Mast Cells
Heart	Mast Cells
Kidney	Mast Cells
Intestine	Mast Cells
Lacrimal gland	Mast Cells
Liver	Mast Cells
Lung	Mast Cells
Parotid gland	Mast Cells
Pancreas	Mast Cells
Peritoneal Cavity	Mature Mast Cells Immature Mast Cells
Skeletal muscle	Mast Cells
Skin	Mast Cells
Spleen	Mast Cells
Sublingual gland	Mast Cells
Submandibular gland	Mast Cells
Testis	Mast Cells
Thymus	Mast Cells
Tongue	Mast Cells

Previous studies using mAb AA4 showed that the α-galactosyl derivatives of the ganglioside GD_{1b} were present only in mast cells and not in any other cell type in all 23 rat tissues examined (Table IV) (Guo *et al.,* 1989; Oliver *et al.,* 1992). However, in bone marrow, a population of large, poorly differentiated cells, presumably immature mast cells also stained with mAb AA4 (Oliver *et al.,* 1992). Later these cells were indeed shown to be very immature and immature mast cells (Jamur *et al.,* 1997 and 2001). Since the heterogeneity of the maturing mast cells makes them impossible to separate from other cells on the basis of their density and mAb AA4 binds only to cells which can be identified as mast cells (Faraco *et al.,* 1997; Jamur *et al.,* 2001), the gangliosides recognized by mAb AA4 may be considered a powerful marker for rodent mast cells.

The ability to characterize the maturation of bone marrow-derived and peritoneal mast cells has been impaired both by the lack of mast cell-specific markers and by the inability to rapidly and efficiently separate mast cells in all stages of maturation from a mixed population of cells (Valent *et al.,* 1991). Using mAb AA4 conjugated to tosylactivated Dynabeads 450, Jamur *et al.,* (1997) successfully separated mast cells from rat bone marrow and peritoneal lavage (Figure 5).

Jamur *et al.,* (2001), then went on to characterize mast cells at various stages of maturation isolated from rat bone marrow. In this study (Jamur *et al.,* 2001), the very immature mast cells, which had not been previously described, contained few granules and would not have been recognized as mast cells by standard cytological methods (Figure 5B). They could however be identified by the presence of the derivatives of the ganglioside GD_{1b} on their surface.

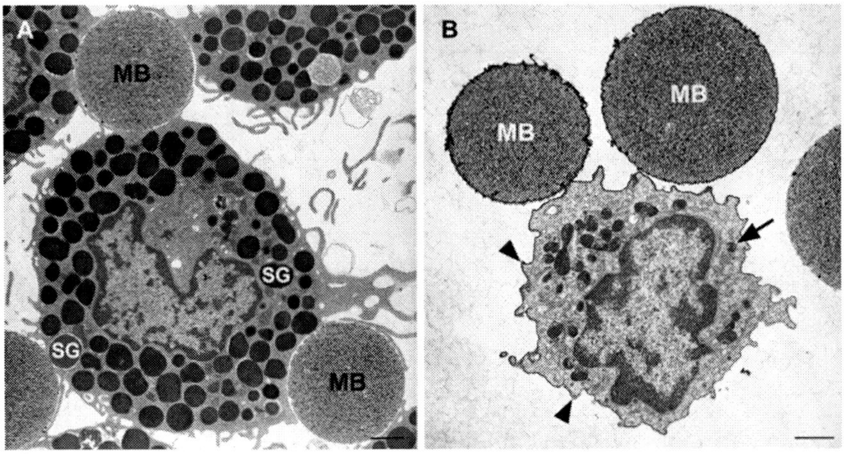

Figure 5. Using mAb AA4 conjugated to magnetic beads, it is possible to isolate mast cells in all stages of maturation. (A) Mature mast cells are bound to mAb AA4 conjugated magnetic beads. (B) A very immature mast cell is seen bound to mAb AA4 conjugated magnetic beads. The cell has very few secretory granules (arrow), but does bind IgE (arrowheads) to its surface. After isolation immature mast cells were immunolabeled with anti-IgE conjugated to horseradish peroxidase to reveal the presence of bound IgE. All samples were processed for transmission electron microscopy. (MB, magnetic bead; SG. secretory granule). Bars equal 1 μm. Micrographs courtesy of Dr. Andrea Moreno, Pontifícia Universidade Católica, Curitiba, Paraná, Brazil.

On the other hand, undifferentiated mast cell precursors in the bone marrow do not express, the α-galactosyl derivatives of the ganglioside GD_{1b} recognized by mAb AA4. The gangliosides begin to be expressed on the cell surface jointly with FcεRI and at the same time as the initiation of the formation of cytoplasmic granules in very immature mast cells and continue to be expressed by mast cells in all stages of maturation (Jamur *et al.,* 2005). The expression of the α-galactosyl derivatives of the ganglioside GD_{1b} on the mast cell surface therefore appears to be related to mast cell maturation.

The α-galactosyl derivatives of the gangliosides GD_{1b} also seem to be intimately involved with signaling through FcεRI. Although the functional role of these gangliosides is not clear, previous studies showed that when the α-galactosyl derivatives of ganglioside GD_{1b} are bound by mAb AA4, histamine release was inhibited in a time and concentration dependent manner and that morphological and biochemical changes similar to those seen with activation of FcεRI were produced (Basciano *et al,* 1986; Oliver *et al.,* 1992; Swaim *et al.,* 1994; Stephan *et al.,* 1997). The morphological changes following activation by mAb AA4 included ruffling of the cell surface, actin polymerization and redistribution of cytoskeletal components (Figure 6). Biochemically, binding of mAb AA4 produced an increase in intracellular calcium, phosphatidylinositol hydrolysis, and a redistribution of PKC. However, the magnitude of these changes was less than after FcεRI aggregation, and unlike FcεRI activation, these changes were not accompanied by histamine release (Oliver *et al.,* 1992).

Minoguchi *et al.* (1994) demonstrated that the α-galactosyl derivatives of the ganglioside GD_{1b} co-precipitated with the Src family tyrosine kinase Lyn and that in spite of the fact that mAb AA4 binds to sites close to FcεRI (Strake *et al.,* 1987; Stephan *et al.,* 1991; Oliver *et al.,* 1992; Swaim *et al.,* 1994) the association between Lyn and these gangliosides was not mediated by FcεRI. The association of Lyn with the gangliosides is much stronger than the association of Lyn with FcεRI. These associations suggest that a complex of molecules that includes gangliosides, FcεRI and Lyn are essential for modulation of signal transduction in mast cells (Minoguchi *et al.,* 1994; Swaim *et al.* 1994).

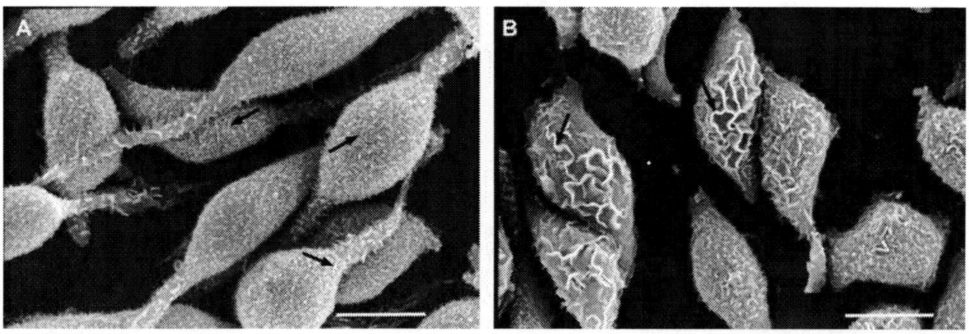

Figure 6. By scanning electron microscopy unstimulated RBL-2H3 cells (A) are elongated and spindle-shaped, and their surface is covered with small, short microvilli (arrows). After stimulation (B), the cells spread on the substratum and their surface is covered with deep membrane ruffles (arrows). Bars equal 10 μm.

Another indication of the involvement of the α-galactosyl derivatives of the ganglioside GD_{1b} with signal transduction in mast cells comes from the finding that FcεRI is co-localized in lipid rafts along with these gangliosides in RBL-2H3 cells (Sheets *et al.*, 1999; Pyenta *et al.*, 2001; Oliver *et al.*, 2007). It is well established that in many cell types, including mast cells, the initial events in signal transduction occur in lipid rafts (Holowka *et al.*, 2000; Wilson *et al.*, 2000; Holowka and Baird, 2001; Rivera *et al.*, 2001b; Pike, 2003; Wilson *et al.*, 2004).

Furthermore, analysis of the subcellular distribution of the gangliosides recognized by mAb AA4 and of FcεRI on sucrose gradients showed that, following FcεRI activation, there was a shift in the distribution of the gangliosides to the lipid raft fractions. The movement of these gangliosides into the lipid rafts may be another mechanism that regulates signal transduction in mast cells. The same study also demonstrated that these mast cell specific gangliosides are internalized in the same vesicles and with the same kinetics as FcεRI (Figure 7). This fact may facilitate the structural preservation of the signaling complexes and the prolongation of the signal (Oliver *et al.*, 2007).

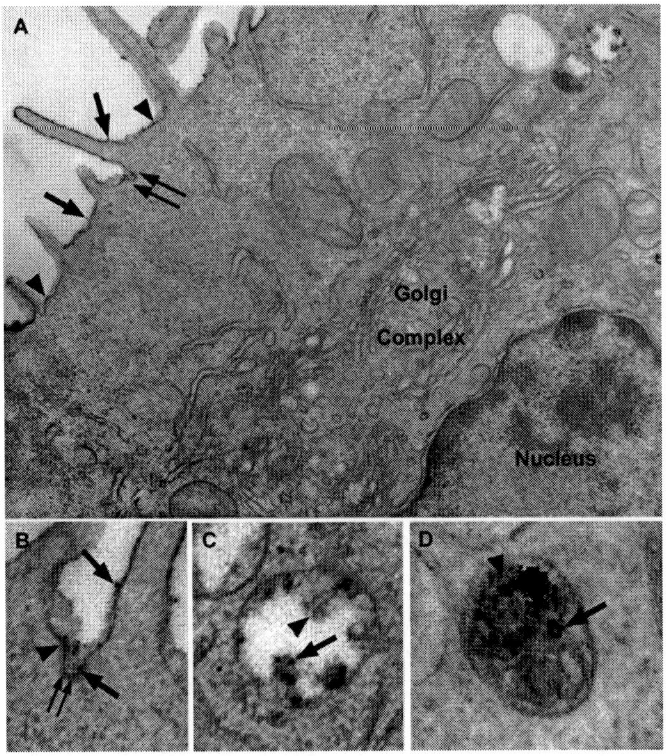

Figure 7. Gangliosides recognized by mAb AA4 are internalized jointly with FcεRI. RBL-2H3 cells were incubated with mAb AA4 conjugated to horseradish peroxidase along with mAb BC4 conjugated to colloidal gold. mAb BC4 recognizes the α-subunit of FcεRI. At varying time intervals, the cells were fixed, incubated in H_2O_2 – diaminobenzidine reaction medium, and processed for transmission electron microscopy. After 5 minutes of incubation (A, B, and C) both the gangliosides and FcεRI can be found in clathrin coated vesicles, double arrows, and early endosomes. At 30 minutes of incubation (D) both the gangliosides and FcεRI can be seen inside a late endosome. Arrows, mAb BC4 gold; arrowheads, mAb AA4-HRP reaction product. Bar A equals 0.5μm, bars B, C and D equal 0.25 μm.

The mutant cell lines (B6A4A2III-E5 and B6A4C1III-D1; a kind gift from Dr. Reuben P. Siraganian, NIDCR-NIH, Bethesda, MD, USA), which are deficient in α-galactosyl derivatives of ganglioside GD_{1b} as well as the parent cell line, RBL-2H3 have also been used to examine the role of these gangliosides in modulating mediator release following stimulation via FcεRI. Compared to the RBL-2H3 cells, the two variants have reduced β-hexosaminidase release after FcεRI stimulation, but not after exposure to calcium ionophore (Silveira e Souza *et al.*, 2004). Field *et al.*, (2000) using another subline of RBL-2H3 cells, also deficient in the gangliosides recognized by mAb AA4, B6A4C1, observed that the absence of gangliosides disturbed FcεRI-mediated degranulation. Stephan *et al.* (1997) also demonstrated, in RBL-2H3 cells, that the anti-ganglioside mAb AA4 selectively inhibits FcεRI-mediated secretion but has no effect on degranulation induced by calcium ionophore. This fact suggests that the association of the derivatives of the ganglioside GD_{1b} and FcεRI is essential for signal transduction through FcεRI and that the binding of mAb AA4 to the gangliosides may impair the function of these gangliosides in signal transduction.

A possible role for the mast cell specific gangliosides in signal transduction is the facilitation of the association of Lyn with FcεRI. Because FcεRI itself has no kinase activity, the tyrosine phosphorylations induced by receptor cross-linking could be secondary to the aggregation of these cell surface complexes (Minoguchi *et al.,* 1994). Therefore, these complexes that include lipid rafts gangliosides, associated kinases, such as Lyn and FcεRI, could be important in receptor-mediated signal transduction.

Furthermore, the α-galactosyl derivatives of the ganglioside GD_{1b} seem to be essential for the maintenance of cell structure and organization in RBL-2H3 mast cells. Also using the two gangliosides deficient cell lines, E5 and D1, and the parent cell line (RBL-2H3), Silveira e Souza *et al.* (2004) observed that the two deficient cell lines were morphologically distinct from each other as well as from the parental RBL-2H3 cells (Figure 8).

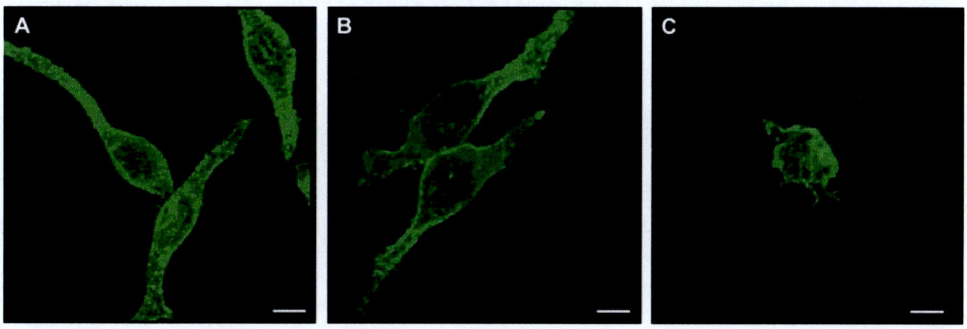

Figure 8. The F-actin distribution, of RBL-2H3, E5 and D1 cell lines reflects their morphology. RBL-2H3 cells and the ganglioside deficient cell lines E5 and D1 were fixed, permeabilized and stained with phalloidin conjugated to Alexa 488. The RBL-2H3 and E5 cells are primarily spindle shaped (A and B, respectively), while the D1 cells are round and smaller (C). Samples were examined using a Leica TCS-NT laser scanning confocal microscope. Bars equal 8 μm.

The RBL-2H3 cells, when stimulated via FcεRI, undergo morphological alterations, loose their normal spindle shaped appearance, increase their ruffling, and spread over the surface of the culture dish (Pfeiffer *et al.*, 1985; Sahara *et al.*, 1990; Oliver *et al.*, 1992). These membrane changes are accompanied by a redistribution of cytoskeletal elements

(Oliver *et al.,* 1992). The E5 cell line, after stimulation through FcεRI had morphological changes similar to the RBL-2H3 cell line, but the pattern of cell surface ruffling was different from that seen with the RBL-2H3 cells. In contrast, the D1 cells did not change their morphology in response to stimulation (Silveira e Souza *et al.,* 2004). The absence of the morphological changes could be related to the membrane lipid composition of D1 cells. In this cell line, there is a large decrease in glycosphingolipids, that could affect many physicochemical properties of the plasma membrane (Kato *et al.,* 2003), as well as the cytoskeletal anchorage to the plasma membrane (Palestini *et al.,* 2000). These dada suggest that the mast cells-specific α-galactosyl derivatives of ganglioside GD_{1b} are also important in maintaining normal cell morphology and the decreased ganglioside expression affects the maintenance of the normal cell structure.

Conclusions

Gangliosides are integral components of lipid rafts. They have roles not only in organizing these microdomains, but also in their signaling functions. The binding specificity of gangliosides as functional ligands depends strictly on their specific structural characteristics and the great structural diversity provides a rich source of structures that can function as binding sites for receptors to mediate and regulate cell signaling pathways. The mast cell-specific α-galactosyl derivatives of the ganglioside GD_{1b} play a role in mast cell maturation, mast cell activation through FcεRI, and in the morphological changes that follow this activation.

Acknowledgments

We would like to thank Mário Traia Castania, for assistance with the graphic representation of lipid rafts, Márcia Sirlene Graeff for assistance with the confocal microscope and José Augusto Maulin for help with the scanning electron microscopy, from the Department of Cell and Molecular Biology and Pathogenic Bioagents, FMRP-USP, Ribeirão Preto, Brazil. The authors would also like to thank CAPES, CNPq, FAEPA, and FAPESP for financial support.

References

Abe, A., Gregory, S., Lee, L., Killen, P.D., Brady, R.O., Kulkarni, A. and Shayman, J.A. (2000). Reduction of globotriaosylceramide in Fabry disease mice by substrate deprivation. *J Clin Invest, 105,* 1563-1571.

Abraham, S.N. and Malaviya, R. (1997). Mast cell in infection and immunity. *Infect Immun, 65*, 3501-3508.

Allende, M.L. and Proia, R.L. (2002). Lubricating cell signaling pathways with gangliosides. *Curr Opin in Struct Biol, 12,* 587-592.

Baird, B., Sheets, E.D. and Holowka, D. (1999). How does the plasma membrane participate in cellular signaling by receptor for immunoglobulin E? *Biophys. Chem., 82,* 109-119.

Basciano, L. K., Berenstein, E. H., Kmak, L. and Siraganian, R. P. (1986).Monoclonal antibodies that inhibit IgE binding. *J. Biol. Chem., 261,* 11823-11831.

Birklé, S., Zeng, G., Gao, L., Yu, R.K. and Aubty, J. (2003). Role of tumor-associated gangliosides in cancer progression. *Biochimie, 85*, 455-463.

Brown, D.A. and Rose, J.K. (1992). Sorting of GPI-anchored proteins to glycolipid-enriched membrane subdomains during transport to the apical cell surface. *Cell, 68,* 533-544.

Brown, D.A. and London, E. (1998). Functions of lipid rafts in biological membranes. *Annu. Rev. Immunol., 15,* 371-404.

Cheng, P.C., Cherukuri, A., Dystra, M., Malapati, S., Sproul, T., Chen, M.R. and Pierce, S.K. (2001). Floating the raft hypothesis: the roles of lipid rafts in B cell antigen receptor function. *Semin. Immunol., 13,* 107-114.

Christie, WW. *Lipid Analysis*. 3rd edition. Bridgwater: Oily Press; 2003.

Dolo, V., D'Ascenzo, S., Sorice, M., Pavan, A., Sciannamblo, M., Prinetti, A., Chigorno, V., Tettamanti, G. and Sonnino, S. (2001). New approaches to the study of sphingolipid enriched membrane domains: the use of electron microscopic autoradiography to reveal metabolically tritium labeled sphingolipids in cell cultures. *Glycoconj J., 17,* 261-268.

Dráber, P. and Dráberová, L. (2001). Lipid rafts in mast cell signaling. *Mol. Immunol., 38,* 1247-1252.

Edidin, M. (2003). Lipids on the frontier: a century of cell-membrane bilayers. *Nat. Rev. Mol. Cell Biol., 5,* 414-418.

Faraco, C., Vugman, I., Siraganian, R.P., Jamur, M.C. and Oliver, C. (1997). Immunocytochemical identification of immature rat peritoneal mast cells using a monoclonal antibody specific for rat mast cells. *Acta Histochem, 99,* 23-27.

Field, K.A., Holowka, D. and Baird, B. (1995). FcεRI-mediated recruitment of $p53/56^{Lyn}$ to detergent-resistant membrane domains accompanies cellular signaling. *Proc. Natl. Acad. Sci. USA, 92,* 9201-9205.

Field, K.A., Apgar, J.R., Hong-Geller, E., Siraganian, R.P., Baird, B. and Holowka, D. (2000). Mutant RBL mast cells defective in Fc epsilon RI signaling and lipid raft biosynthesis are reconstituted by activated Rho-family GTPases. *Mol. Biol. Cell, 10,* 3661-3673.

Fishman, P.H. (1986). Recent advances in identifying the functions of gangliosides. *Chem. Phys. Lipides, 42,* 137-151.

Fujinaga, Y., Wolf, A.A., Rodighiero, C., Wheeler, H., Tsai, B., Allen, L., Jobling, M.G., Rapoport, T., Holmes, R.K. and Lencer, W. (2003). Gangliosides that associated with lipid rafts mediate transport of cholera and related toxins from the plasma membrane to endoplasmic reticulum. *Mol. Biol. Cell, 14*, 4738-4793.

Galli, S.J. (2000). Mast cells and basophils. *Curr. Opin. Hematol., 7*, 32-39.

Galli, S.J., Kaleniskoff, J., Grimbaldeston, M.A., Piliponsky, A.M., Williams, C.M.M. and Tsai, M. (2005). Mast cell as "tunable" effector and immunoregulatory cells. *Annu. Rev. Immunol., 23,* 749-786.

Gammon, C.M. and Ledeen, R.W. (1985). Evidence for the presence of a ganglioside transfer protein in brain. *J. Neurochem., 44*, 979-982.

Gillard, B. K., Thurmon, L.T. and Marcus, D.M. (1993). Variable subcellular localization of glycosphingolipids. *Glycobiology, 3*, 57-67.

Gilfillan, A. and Tkaczyk, C. (2006). Integrated signalling pathways for mast-cell activation. *Nat. Rev. Immunol., 6*, 218-230.

Guo, N., Her, G. R., Reinhold, V. N., Brennam, M. J., Siraganian, R. P. and Ginsburg, V. (1989). Monoclonal antibody AA4 which inhibits binding of IgE to high affinity receptors on rat basophilic leukemia cells, binds to novel α-galactosyl derivates of ganglioside GD_{1b}. *J. Biol. Chem., 264,* 13267-13272.

Hakomori, S. (1981). Glycosphingolipids in cellular interaction, differentiation, and oncogenesis. *Annu. Rev. Biochem., 50*, 733-764.

Hakomori, S. (1990). Bifunctional role of glycosphingolipids. *J. Biol. Chem., 265*, 18713-18716.

Hakomori, S. (1993). Structure and function of sphingolipids in transmembrane signaling and cell-cell interaction. *Biochem. Soc. Trans, 21*, 583-595.

Hakomori, S. (1996). Tumor malignancy defined by aberrant glycosylation and sphingo(glyco)lipid metabolism. *Cancer Res., 56*, 5309-5318.

Hakomori, S. (2002). Inaugural article: the glycosynapse. *Proc. Natl. Acad. Sci. USA, 99*, 225-232.

Hannun, Y.A. and Bell, R.M. (1989). Functions of sphingolipids and sphingolipids breakdown products in cellular regulation. *Science, 243*, 500-507.

Harder, T. (2001). Raft membrane domains and immunoreceptor functions. *Adv. Immunol., 77*, 45- 92.

Hideharu, I. and Makoto, K. (2001). Systematic Synthesis of Gangliosides. Trends in *Glycoscience and Glycotechnology, 13*, 57-64.

Holowka, D., Gosse, J.A., Hammond, A.T., Han, X., Sengupta, P., Smith, N.L., Wagenknecht-Wiesner, A., Wu, M., Young, R.M., and Baird B. (2005). Lipid segregation and IgE receptor signaling: A decade of progress. *Biochim. Biophys. Acta, 20,* 1-8.

Holowka, D., and Baird B. (2001). FcεRI as a paradigm for a lipid raft-dependent receptor in hematopoietic cells. *Semin. Immunol., 13*, 99-105.

Holowka, D., Sheets, E.D. and Baird, B. (2000). Interactions between FcεRI and lipid rafts components are regulated by actin cytoskeleton. *J. Cell Sci., 133,* 1009-1019.

Hunter, T. (2000). Signaling-2000 and beyond. *Cell, 100,* 113-127.

Igarashi, Y., Nojiri, N., Hanai, N. and Hakomori, S. (1989). Gangliosides that modulate membrane protein function. *Methods Enzym., 179,* 521-541.

IUPAC-IUB - Joint Commision on Biochemical Nomenclatures (JCBN). Nomenclature of glycolipids. Recommendations 1997. (1998). *Eur. J. Biochem., 257*, 293-298.

Iwabuchi, K., Yamamura, S., Prinetti, A., Handa, K. and Hakomori, S. (1998). GM3-enriched microdomain involved in cell adhesion and signal transduction through carbohydrate-carbohydrate interaction in mouse melanoma B16 cells. *J. Biol. Chem., 15,* 9130-9138.

Jamur, M.C., Grodzki, A.C.G., Moreno, A.N., Swaim, W., Siraganian, R.P. and Oliver, C. (1997). Immunomagnetic isolation of rat bone marrow-derived and peritoneal mast cells. *J. Histochem. Cytochem., 45,* 1715-1722.

Jamur, M.C., Grodzki, A.C.G., Moreno, A.N., de Mello, L.F., Pastor, M.V.D., Berenstein, E.H., Siraganian, R.P. and Oliver, C. (2001). Identification and isolation of rat bone marrow-derived mast cells using the mast cell-specific monoclonal antibody AA4. *J. Histochem. Cytochem., 49*, 219-228.

Jamur, M.C., Grodzki, A.C.G., Berenstein, E.H., Hamawy, M.M., Siraganian, R.P. and Oliver, C. (2005). Identification and characterization of undifferentiated mast cells in mouse bone marrow. *Blood, 105,* 4282-4289.

Kato, N., Nakanishi, M. and Hirashima, N. (2003). Cholesterol depletion inhibits store-operated calcium currents and exocytotic membrane fusion in RBL-2H3 cells. *Biochemistry, 42,* 11808-11814.

Kimura, T., Hisano, M., Inoue, Y. and Adachi, M. (2001). Tyrosine phosphotylation of the linker for activator T cells in mast cells by stimulation with the high affinity IgE receptor. *Immunol. Lett., 75*, 123-129.

Kinet, J. P. (1999). The high-affinity IgE receptor (FcεRI): from physiology to pathology. *Annu. Rev. Immunol., 17,* 931-972.

King, M.W. and Marchesini, S. Metabolism of the Sphingolipids. 2002 February 14. Available from: http://www.med.unids.it/~marchesi/lipsynth3.html. Search for "Lipid metabolism" from the opening page.

Kitamura, Y. (1989). Heterogeneity of mast cell and phenotypic change between subpopulations. *Annu. Rev. Immunol., 7,* 59-76.

Knoll, F., Kolter, T. and Sandhoff, K. (2000). Sphingolipid photoaffinity labels. *Methods Enzymol, 311*: 568-576.

Kolter, T., Proia, R.L. and Sandhoff, K. (2002). Combinatorial ganglioside biosynthesis. *J. Biol. Chem., 277*, 25859-25862.

Li, R., Manela, J., Kong, Y. and Ladish, S. (2000). Cellular gangliosides promote growth factor-induced proliferation of fibroblasts. *J. Biol. Chem., 275,* 34213-34223.

Liu, Y., Wada, R., Kawai, H., Sango, K., Deng, C., Tai, T., McDonald, M.P., Araújo, K., Crawley, J.N., Bierfreund, U., Sandhoff, K., Suzuki, K. and Proia, R.L. (1999). A genetic model of substrate deprivation therapy for a glycosphingolipid storage disorder. *J. Clin. Invest, 103,* 497-505.

Maccioni, H. J. F., Daniotti, J. L. and Martina, J. A. (1999). Organization of ganglioside synthesis in Golgi apparatus. *Biochim. Biophys. Acta, 1437*: 101-118.

Melkonian, K.A., Ostermeyer, A.G., Chen, J.Z., Roth, M.G. and Brown, D. (1999). Role of lipid modifications in targeting proteins to detergent-resistant membrane rafts. *J. Biol. Chem., 274,* 3910-3917.

Metcalfe, D.D., Baram, D. and Mekori, Y.A. (1997). Mast cells. *Physiol Rev.,* 77, 1033-1079.

Metelmann, W., Vukelić, Ž. and Peter-Katalinić, J. (2001). Nano-electrospray ionization time-of-flight mass spedtrometry of gangliosides from human brain tissue. *J. Mass Spectrom., 36*, 21-29.

Metzger, H. (1992). The receptor with high affinity for IgE. *Immunol Rev., 125,* 37-48.

Metzger, H. (1999). It's spring, and thoughts turn to…allergies. *Cell, 97,* 287-290.

Miceli, M.C., Moran, M., Chung, C.D., Patel, V.P., Low, T. and Zinnanti, W. (2001). Co-stimulation and counter-stimulation: lipid raft clustering controls TCR signaling and functional outcomes. *Semin. Immunol., 13,* 115-128.

Meuillet, E.J., Mania-Farnell, B., George, D., Inokuchi, J.I. and Bremer, E.G. (2000). Modulation of EGF receptor activity by changes in the GM3 content in a human epidermoid carcinoma cell line, A431. *Exp. Cell Res., 256,* 74-82.

Miljan, E.A., Meuillet, E.J., Mania-Farnell, B., George, D., Yamamoto, H., Simon, H.G. and Bremer, E.G. (2002). Interaction of the extracellular domain of the epidermal growth factor receptor with gangliosides. *J. Biol. Chem., 277,* 10108-10113.

Minoguchi, K., Swaim W.D., Berenstein, E.H. and Siraganian, R.P. (1994). Src family tyrosine kinase $p53/56^{Lyn}$, a serine kinase and Fc epsilon RI associate with alpha-galactosyl derivatives of ganglioside GD_{1b} in rat basophilic leukemia RBL-2H3 cells. *J. Biol. Chem. 269,* 5249-5254.

Möbius, W., Herzog, V., Sandhoff, K. and Schwarzmann, G. (1999). Gangliosides are transported from the plasma membrane to intralysosomal membranes as revealed by immuno-electron microscopy. *Biosci. Rep., 19,* 307-316.

Nadler, M.J., Mattheus, S.A., Turner, H. and Kinet, J.P. (2000). Signal transduction by the high-affinity immunoglobulin E receptor FcεRI: coupling form to function. *Adv. Immunol., 76,* 325-355.

Nelson, DL; Cox, MM. Lipids. In: *Lehninger Principles of Biochemistry*. New York, NY: Worth Publishers; 2000; 363-388.

Oliver, C., Sahara, N., Kitani, S., Robbins, A.R., Mertz, L.M. and Siraganian, R.P. (1992). Binding of monoclonal antybody AA4 to gangliosides on rat basophilic leukemia cells produces changes similar to those seen with Fcepsilon receptor activation. *J. Cell Biol., 116,* 635-646.

Oliver, C., Fujimura, A., Silveira e Souza, A.M.M., Orlandini de Castro, R., Siraganian, J.P. and Jamur, M.C. (2007). Mast cell-specific gangliosides and FcepsilonRI follow the same endocytic pathway from lipid rafts in RBL-2H3 cells. *J. Histochem. Cytochem., 55,* 315-25.

Osterbye, T., Jorgensen, K.H., Fredman, P., Tranum-Jensen, J., Kaas, A., Brange, J., Whittingham, J.L. and Buschard, K. (2001). Sulfatide promotes the folding of proinsulin, preserves insulin crystals, and mediates its monomerization. *Glycobiology, 11,* 473-479.

Palestini, P., Pitto, M., Tedeschi, G., Ferraretto, A., Parenti, M., Brunner, J. and Masserini, M. (2000). Tubulin anchoring to glycolipid-enriched, detergent-resistant domains of the neuronal plasma membrane. *J. Biol. Chem., 276,* 9978-9985.

Parton, R.G. and Richards, A.A. (2003). Lipid rafts and caveolae as portals for endocytosis: new insights and common mechanism. *Traffic, 4,* 724-738.

Pfeiffer, J. R., Seagrave, J. C., Davis, B. H., Deanin, G. G. and Oliver, J. M. (1985). Membrane and cytoskeletal changes associated with IgE-mediated serotonin release from rat basophilic leukemia cells. *J. Cell Biol., 101,* 2145-2155.

Pike, L.J. (2003). Lipid rafts: bringing order to chaos. *J. Lipid Res., 44,* 655-667.

Pike, L.J. (2004). Lipid Rafts: heterogeneity on the high seas. *Biochem J., 378,* 281-292.

Platt, F.M., Neises, G.R., Reinkensmeier, G., Townsend, M.J., Perry, V.H., Proia, R.L., Winchester, B., Dwek, R.A. and Butters, T.D. (1997). Prevention of lysosomal storage in Tay-Sachs mice treated with *N*-butyldeoxynojirimycin. *Science, 276,* 428-431.

Pralle, A., Keller, P., Florin, E.L., Simons, K. and Horber, J.K. (2000). Sphingolipid-cholesterol rafts diffuse as small entities in the plasma membrane of mammalian cells. *J. Cell Biol., 148*, 997-1008.

Pribluda, V.S., Pribluda, C. and Metzger, H. (1994). Transphosphorylation as the mechanism by which the high-affinity receptor for IgE is phosphorylated upon aggregation. *Proc. Natl. Acad. Sci. USA*, 91, 11246-11250.

Pomorski, T., Hrafnsdóttir, S., Devaux, P. and van Meer, G. (2001). Lipid distribution and transport across cellular membranes. *Semin. Cell Dev. Biol., 12*, 139-148.

Pyenta, P.S., Holowka, D. and Baird, B. (2001). Cross-correlation analysis of inner-leaflet-anchored green fluorescent protein co-dedistributed with IgE receptors and leaflet lipid raft components. *Biophys J., 80,* 2120-2132.

Radin, N. S. (1999). Chemotherapy by Slowing Glucosphingolipid Synthesis. *Biochem. Pharmacol., 57,* 589-595.

Rajendran, L. and Simons, K. (2005). Lipid rafts and membrane dynamics. *J. Cell Sci., 118,* 1099-1102.

Rebbaa, A., Hurth, J., Yamamoto, H., Kersey, D.S. and Bremer, E.G. (1996). Ganglioside GM_3 inhibition of EGF receptor mediated signal transduction. *Glycobiology, 6,* 399-406.

Reynolds, C.P., Maurer, B.J. and Kolesnick, R.N. (2004). Ceramide synthesis and metabolism as a target for cancer therapy. *Cancer Lett., 206,* 169-180. 37-141.

Rivera, J., Cordero, J.R., Furumoto, Y., Luciano-Montalvo, C., Gonzalez-Espinosa, C., Kovarova, M., Odom, S. and Parravicini, V. (2001a). Macromolecular protein signaling complexes and mast cell responses: a view of the organization of IgE dependent mast cell signaling. *Mol. Immunol., 38,* 1253-1258.

Rivera, J., Arudchandran, R., Gonzalez-Espinosa, C., Manetz, T.S. and Xirasagar, S. (2001b). A perspective: regulation of IgE receptor-mediated mast cell responses by a LAT-organized plasma membrane-localized signaling complex. *Int. Arch. Allergy Immunol., 124,* 137-141.

Rivera, J., Gonzales-Espinosa, C., Kovarova, M. and Parravicini, V. (2002). The architecture of IgE-dependent mast cell signaling. *ACI International, 14*, 25-36.

Robbie-Ryan, M. and Brown, M. (2002). The role of mast cells in allergy and immunity. *Curr. Opin. Immunol., 14,* 728-733.

Rodgers, W. and Rose, J.K. (1996). Exclusion of CD45 inhibits activity of p56lck associated with glycolipid-enriched membrane domains. *J. Cell Biol., 135,* 1515-1523.

Sahara, N., Siraganian, R.P. and Oliver, C. (1990). Morphological changes induced by the calcium ionophore A23187 in rat basophilic leukemia (2H3) cells. *J. Histochem and Cytochem., 38*, 975-983.

Sheets, E.D., Holowka, D. and Baird, B. (1999). Membrane organization in immunoglobulin E receptor signaling. *Curr. Opin. Chem. Biol., 3*, 95-99.

Silveira e Souza, A.M.M., Jamur, M. C., Siraganian, R.P. and Oliver, C. (2004). Characterization of RBL-2H3 mast cell variants deficient in the gangliosides recognized by mAb AA4. In: 44th Annual Meeting of the American Society for Cell Biology, Washington – DC, *Mol. Biol. Cell, 15*: 627a-627a.

Simons, K. and Ikonen, E. (1997). Functional rafts in cell membranes. *Nature, 387*, 569-572,

Simons, K. and Toomre, D. (2000). Lipid rafts and signal transduction. *Nat. Rev. Mol. Cell Biol., 1*, 31-41.

Singer, S.J. and Nicolson, G.L. (1972). The fluid mosaic model of the structure of cell membranes. *Cell, 175*: 720-731.

Smart, E.J., Ying, Y., Mineo, C. and Anderson, R.G.W. (1995). A detergent-free method for purifying caveolae membrane from tissue culture cells. *Proc. Natl. Acad. Sci. USA, 22,* 10104-10108.

Siraganian, RP. Mast cells and basophils. In Gallin, J.J., Goldstein, I.M., Snyderman, R., eds. *Inflamation: Basic Principles and Clinical Correlates.* New York, NY: Raven Press; 1988; 513-542.

Siraganian, R.P. (2003). Mast cell signal transduction from the high-affinity IgE receptor, *Curr. Opin. Immunol., 15,* 639-646.

Sixma, T.K., Kalk, K.H., van Zanten, B.A., Dauter, Z., Kingma, J., Wiyholt, B. and Hol, W.G. (1993). Refined structure of Escherichia coli heat-labile entero-toxin, a close relative of cholera toxin. *J. Mol. Biol., 230*, 890-918.

Stauffer, T.P. and Meyer, T. (1997). Compartimentalized IgE receptor-mediated signal transduction in living cell. *J. Cell Biol., 139*, 1447-1454.

Stephan, V., Guo, N., Ginsburg, V. and Siraganian, R.P. (1991). Immunoprecipitation of membrane proteins from rat basophilic leukemia cells by the antiganglioside monoclonal antibody AA4. *J. Immunol., 146*, 4271-4277.

Stephan, V., Seibt, A., Dukanovic, D., Skasa, M., Swaim, W.D., Berenstein, E.H., Siraganian R.P. and Wahn, V. (1997). Anti-ganglioside monoclonal antibody AA4 selectively inhibits IgE-induced signal transduction pathways in rat basophilic leukemia cells. *Mol. Immunol., 34*, 227-235.

Stracke, M.L., Basciano, L.K. and Siraganian, R.P. (1987). Binding properties and histamine release in variants of rat basophilic leukemia cells with changes in the IgE receptor. *Immunol Lett, 14,* 287-292.

Stump, R.F., Pfeiffer, J.R., Schneebeck, M.C., Seagrave, J.C. and Oliver, J.M. (1989). Mapping gold-labeled receptors on cell surfaces by backscattered electron imaging and digital image analysis: studies of the IgE receptor on mast cells. *Am. J. Anat., 185,* 128-141.

Svennerholm, L. (1963). Chromatographic separation of human brain gangliosides. *J. Neurochem.,* 10:613-623.

Svennerholm, L. (1980). Ganglioside designation. *Adv. Exp. Med. Biol., 125,* 11-19.

Svennerholm, L., Böstrom, K., Fredman, P., Manson, J.E. and Rynmark, B.M. (1989). Human brain gangliosides: developmental changes from early fetal stage to advanced age. *Biochim. Biophys. Acta, 1005*, 109-117.

Svennerholm, L., Rynmark, B.M., Vibergsson, G., Fredman, P., Gotfries, J., Manson, J.E. and Percy, A.J. (1991). Gangliosides in human fetal brain. *J. Neurochem., 56*, 1763-1768.

Swaim, W.D., Minoguchi, K., Oliver, C., Hamawy, M.M., Kihara, H., Stephan, V., Berenstein, E.H. and Siraganian, R.P. (1994). The anti-ganglioside monoclonal antibody AA4 induces protein tyrosine phosphorylations, but not degranulation, in rat basophilic leukemia cells. *J. Biol. Chem., 269*, 19466-19473.

Tagami, S., Inokuchi, Ji J., Kabayama, K., Yoshimura, H., Kitamura, F., Uemura, S., Ogawa, C., Ishii, A., Saito, M., Ohtsuka,Y., Sakaue, S. and Igarashi, Y. (2002). Ganglioside GM_3 participates in the pathological conditions of insulin resistance. *J. Biol. Chem., 277,* 3085-92.

Tifft, C.J. and Proia, R.L. (2000). Stemming the tide: glycosphingolipids synthesis inhibitors as therapy for storage diseases. *Glycobiology, 10,* 1249-1258.

Thompson, T.E. and Tillack, T.W. (1985). Organization of glycosphingolipids in bilayers and plasma membranes of mammalian cells. *Ann. Rev. Biophys. Chem. 14*, 361-39-86.

Valent, P., Silaber, C. and Betterlhein, P. (1991). The growth and differentiation of mast cells. *Prog. Growth Factor Res., 3,* 27-41.

Varma, R. and Mayor, S. (1998). GPI-anchored proteins are organized in subdomains at the cell surface. *Nature, 394,* 798-801.

Varki, A., Cummings, R., Esko, J., Hart, G., and Marth, J. (1999). *Essential of Glycobiology* (1st edition). La Jolla, CA. Consortium of Glycobiology editors.

Xavier, R., Brennan, T., Li, Q., McCormack, C. and Seed, B. (1998). Membrane compartmentation is required for efficient T cell activation. *Immunity, 8,* 723-732.

Wang, X.Q., Sun, P., O'Gorman, M., Tai, T. and Paller, A.S. (2001). Epidermal growth factor receptor glycosylation is required for ganglioside GM_3 binding and GM_3-mediated suppression of activation. *Glycobiology, 11,* 515-522.

Werth, N., Schuette, C.G., Wilkening, G., Lemn, T. and Sandhoff, K. (2001). Degradation of membrane-bound ganglioside GM_2 by β-hexosaminidase A. Stimulation by GM_2 activator protein and lysosomal lipids. *J. Biol. Chem., 276,* 12685-12690.

Williams, C.M. and Galli, S.J. (2000). The diverse potential effector and immunoregulatory roles of mast cells in allergic disease. *J. Allergy Clin. Immunol., 105*, 847-859.

Wilson, B.S.; Pfeiffer, J.R. and Oliver, J.M. (2000). Observing FcεRI signaling from the inside of the mast cell membrane. *J. Cell Biol., 149*, 1131-1142.

Wilson, B.S.; Pfeiffer, J.R., Surviladze, Z., Gaudet, E.A., and Oliver, J.M. (2001). High resolution mapping reveals distinct FcεRI and LAT domains in activated mast cells. *J. Cell Biol., 145,* 645-658.

Wilson, B.S.; Steinberg, S.L., Liederman, K., Pfeiffer, J.R., Surviladze, Z., Zhang, J., Samelson, L.E., Yang, L., Kotula, P.G. and Oliver, J.M. (2004). Markers for detergent-resistant lipid rafts occupy distinct and dynamic domains in native membranes. *Mol. Biol. Cell, 12,* 2580-2592.

Yamashita, T., Wada, R., Sasaki, T., Deng, C., Bierfreund, U., Sandhoff, K. and Proia, R.L. (1999). A vital role for glycosphingolipid synthesis during development and differentiation. *Proc. Natl. Acad. Sci. USA, 96*, 9142-9147.

Yong, L. C. J. (1997). The mast cells: origin, morphology, distribution, and function. *Exp. Toxic Pathol., 49*, 409-424.

Young, R.M., Holowka, D. and Baird, B. (2003). A lipid raft environment enhances Lyn kinase activity by protecting the active site tyrosine from dephosphorylation. *J. Biol. Chem., 278*, 20746–20752.

Yu, R.K. (1994). Development regulation of ganglioside metabolism. *Prog. Brain Res., 101,* 31-44.

Zhang, R., Scott, G., Westbrook, D.L., Nance, M.L., Spangler, B.D., Shipley, G.G. and Westbrook, E.M. (1995). The three-dimensional crystal structure of cholera toxin. *J. Mol. Biol., 251,* 563-573.

In: Glycolipids: New Research
Editor: Daiki Sasaki, pp. 79-98

ISBN: 978-1-60456-216-3

Chapter III

Glycolipids: Biomarkers for Neural Stem and Progenitor Cells

Perrine Barraud[1] and Simon RW Stott[2]
[1]Neurosciences, Veterinary Medicine, Madingley Road,
Cambridge, CB30ES, UK
[2]The National Institute for Medical Research,
The Ridgeway, Mill Hill London, NW7 1AA, UK

Abstract

Glycolipids are, by definition, long carbohydrate and hydrophobic molecules that integrate into the plasma membrane and bind to diverse types of sugar residues at their extracellular domain. Their location on the external surface of the cell makes them excellent biomarkers for the prospective isolation of various subsets of cells from the central nervous system. In this chapter, we review glycolipids that have been used to identify diverse populations of neural stem and progenitor cells during development and in the adult. In particular, we focus our discussion on *stage specific embryonic antigen 1* and *4* (SSEA-1 and SSEA-4, respectively). These two glycolipids are widely used to monitor embryonic stem cell differentiation, but have more recently been employed in the identification and positive selection of two distinct subsets of neural stem and progenitor cells. We explore their possible biological functions and discuss differences between species. And finally, we speculate on their possible applications in the therapeutic correction of neurological damage.

Introduction

Biological membranes from different cells have very similar compositions. The general structure consists of a bi-layer of amphipathic lipids that spontaneously organize themselves so that their hydrophobic "tails" are protected from the surrounding fluids. This arrangement encases the contents of the cell and allows the cell to tightly regulate what enters it, as well as preventing the passive diffusion of hydrophilic molecules. Upon an inspection of the cell surface at the microscopic level, one would observe that the outer envelope consists predominantly of lipids and proteins. There are 3 kinds of lipids in the cell membrane according to the general literature: phospholipids, cholesterol/steroids and glycolipids. Of these, glycolipids are striking due to their asymmetric distribution. These molecules are found almost exclusively in the non-cytoplasmic half of the lipid bilayer. A great deal is now known about the structure, synthesis and expression pattern of these glycolipids. But given their expression on the external cell surface of diverse cell types in the central nervous system (CNS), they represent valuable biomarkers in many fields of scientific investigation and therapeutic medicine. In this chapter, we shall discuss what glycolipids are specifically expressed in the developing and adult CNS, and their recent utility in isolating specific subsets of neural stem and progenitor cells.

1. Glycolipids

Glycolipids can be considered part of a larger family known as glycoconjugates. These are molecules in the cell surface that chains of carbohydrates bind to. In addition to glycolipids, other major types of glycoconjugates include glycoproteins, glycopeptides, peptidoglycans, proteoglycans, and lipopolysaccharides. Glycoconjugates have many biological functions in the body, but are primarily believed to be involved in cell-cell recognition and interactions [1]. All of these glycoconjugates most likely perform fundamental roles, but here we will focus specifically on glycolipids.

As the name suggests, the basic structure of a glycolipid consists of a carbohydrate attached to a lipid. The lipid portion of glycolipids is ceramide, which consists of fatty acids of varying lengths that bind to a long-chain of aliphatic amino alcohol called sphingosine [2]. It should be noted that phospholipids are also derived from ceramide, but glycolipids differ in that one or more sugars attach to the primary hydroxyl group of sphingosine (as opposed to phosphoryl choline in the case of phospholipids). The carbohydrate portion of the glycolipid is a rather more complex issue and the chains of sugar residue that attach to the ceramide differ between subsets of these cell surface molecules. The differences in carbohydrates results in various types of glycolipids, but the majority of them are referred to as "Glycosphingolipids" (GSLs, also known as "Sphingoglycolipids").

2. Synthesis

The plasma membrane is an extremely dynamic environment, with a high level of internalisation and recycling - even in non-dividing cells. It has been suggested that at least half of the cell membrane in fibroblasts is replaced every hour [3]. Given that glycolipids make up a significant proportion of the cell surface – 20-25% of the outer cell membrane of neurons [4] - a rapid production of these molecules is required. Glycolipid synthesis takes place on intracellular membranes and is initiated by membrane bound enzymes. A simplified schematic representation of the synthesis of glycolipids is presented in figure 1.

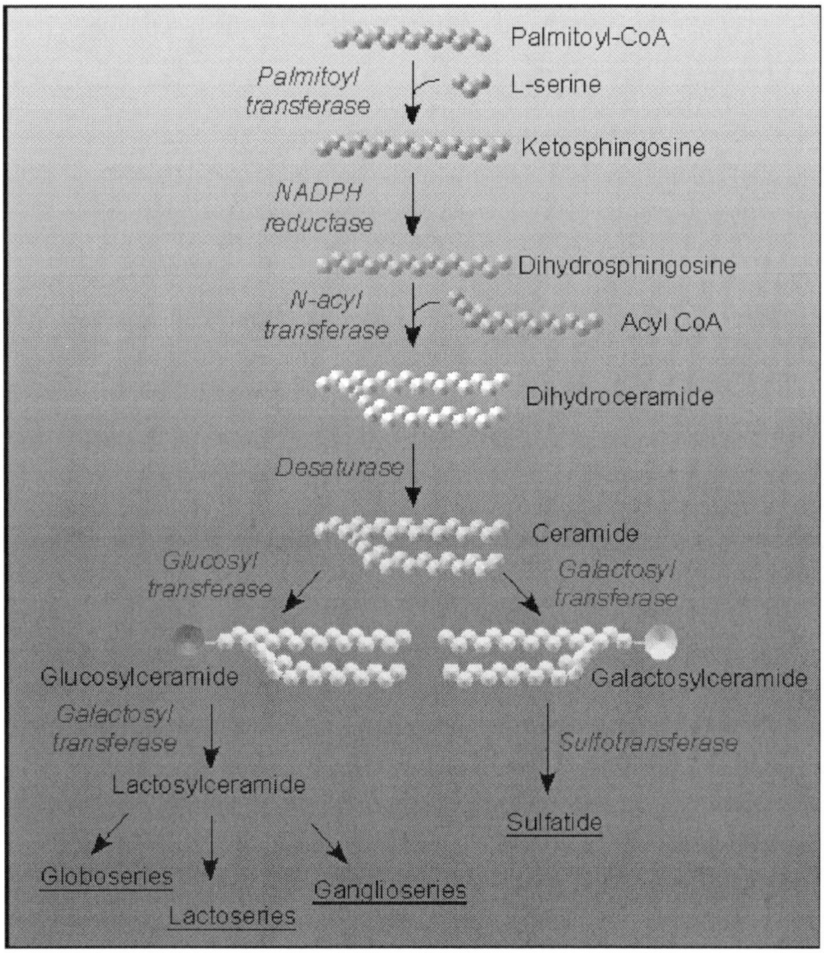

Figure 1. Schematic representation of the biosynthesis of sulfatides, globo-, lacto- and ganglio-series.

It begins with the condensing of activated palmitic acid (palmitoyl-co-enzyme A) and serine to form ketosphingosine. This reaction is initiated by palmitoyl transferase in the endoplasmic reticulum. Ketosphingosine is further reduced, via a NADPH-dependent reductase, into the long-chain base, dihydrosphingosine. Dihydrosphingosine does not occur freely in cells as it is immediately converted into dihydroceramide by an acyltransferase which acylates a long-chain fatty acid (acyl CoA) with the 2-amino group of the sphingosine. Dihydroceramide desaturase then introduces a double bond, which acts only on sphingosine, giving rise to ceramide (N-acyl sphingamide) [4]. Ceramide can also be formed by recycling via the hydrolysis of sphingomyelin by sphingomyelinase.

As stated above, ceramide constitutes the foundation of glycolipids to which chains of sugar residues attach. With regards to glycolipids, ceramide can be either galactosylated or glucosylated to form galactosylceramide (galacto-cerebroside) and glucosylceramide (gluco-cerebroside), respectively (see figure 1). This process takes place in the golgi membrane.

Although galactosylceramide is not often glycosylated, it is frequently sulfated by sulfotransferase resulting in sulfatide. When compared with other groups of glycolipids, there are relatively few sulfated glycolipids. The most abundant is sulfatide, which is a major lipid component of myelin in the white matter of the CNS [5]. The first step in the glycosylation of glucosylceramide is the generation of lactosylceramide, by galactosyltransferase. Lactosylceramide is the precursor of all GSLs. From this particular glycolipid, three distinct families of GSLs are derived: Ganglio-series, Globo-series, and Lacto-series. Ganglio-series glycolipids (gangliosides) are GSLs that contain sialic acid. Globo-series glycolipids differ from ganglio-series, in that their production begins with galactosylation of lactosylceramide. The third group, lacto-series GSLs, begin with the transfer of N-acetylglucosamine to lactosylceramide.

Within these three major subsets of GSLs reside numerous combinations of glycolipids. Each has differing expression patterns for defined subsets of cells (both immature and differentiated) and as a result they represent biomarkers that can be used in different experimental settings. In the rest of this chapter, we shall focus on the glycolipids that have been identified based on their expression in the proliferative regions of the brain, and consequently used for the prospective isolation of subsets of neural stem/progenitor cells. In addition, we will look at recent advances that offer perspective for the translation of stem cell biology to stem cell therapy.

Glycolipids Expressed by Neural Stem Cells

Several classes of glycolipids have been identified in the developing and adult CNS of both human and rodent, but their mode of expression varies among vertebrate species. While some types of glycolipids are particularly expressed by subsets of neuronal or glial cells, others are found to be associated with subpopulations of neural stem cells, such as the lacto-series associated glycolipid SSEA-1, the globo-series associated glycolipids SSEA-3 and SSEA-4, and the gangliosides G_{M1}, G_{D3} and G_{T3}.

1. The Stage Specific Embryonic Antigens (SSEAs)

In the seventies, it was proposed that mechanisms involved in controlling development and differentiation of the early mouse embryo prior to its implantation (the "embryonic stem cells" or ES cells) were likely to be governed by cell surface glycoconjugates [6-9]. Based on this assumption, monoclonal antibodies were developed to further explore such cell surface antigens. These studies used teratocarcinoma cells (originating from abnormal proliferation of male germ cells which develop into structures morphologically similar to normal embryonic structures like morulae and blastulae) as immunogens [10]. Immunohistochemistry on mouse ES cells and also during early development of the mouse embryo not only demonstrated that such antibodies recognise mouse ES cells, but that they also identified distinct glycolipid-related antigens defined as stage specific embryonic antigens SSEA-1 (and its related LewisX antigen), SSEA-3 and SSEA-4 [10,11].

1.1. The Stage Specific Embryonic Antigen 1

The anti-SSEA1 antibody recognizes a lacto-series (see illustration in figure 2; adapted from [11,12]) associated with a carbohydrate core structure. Anti-SSEA-1 also binds to an oligosaccharide containing the Lewis X or Le^x, an epitope which is also known as leucocyte cluster of differentiation 15 (CD15) [13]. Both SSEA-1/CD15 and Le^x epitope comprised the same structure [Galβ1→4(Fucα1→3)GlcNAc-R] [13,14]. Furthermore, the anti-FORSE-1 antibody has been shown to recognize Le^x-containing glycoconjugates [15]. In this chapter we will use the term SSEA-1 for molecules related to SSEA-1 like Le^x or CD15 for simplification, even if molecular structures between SSEA-1 and CD15 are not identical. In addition we will only present data obtained from rodents and humans.

The SSEA-1 antigen, which has primarily been identified on undifferentiated mouse ES cells but not on human ES cells, was also detected in both the developing and the adult mouse CNS [16-20]. In vertebrates, the nervous system develops in two successive steps: a neurogenic one in which only neuroblasts within the neuroepithelium proliferate. This neurogenic phase is followed by a gliogenic step that is characterized by appearance of astrocytes and oligodendrocytes (for review see [21]). In the mouse embryo (at embryonic day (E)8 to 10), immunoreactivity for SSEA-1 was observed as early as E8 in the neuroepithelium and later in the neural tube at E10 [18,22].

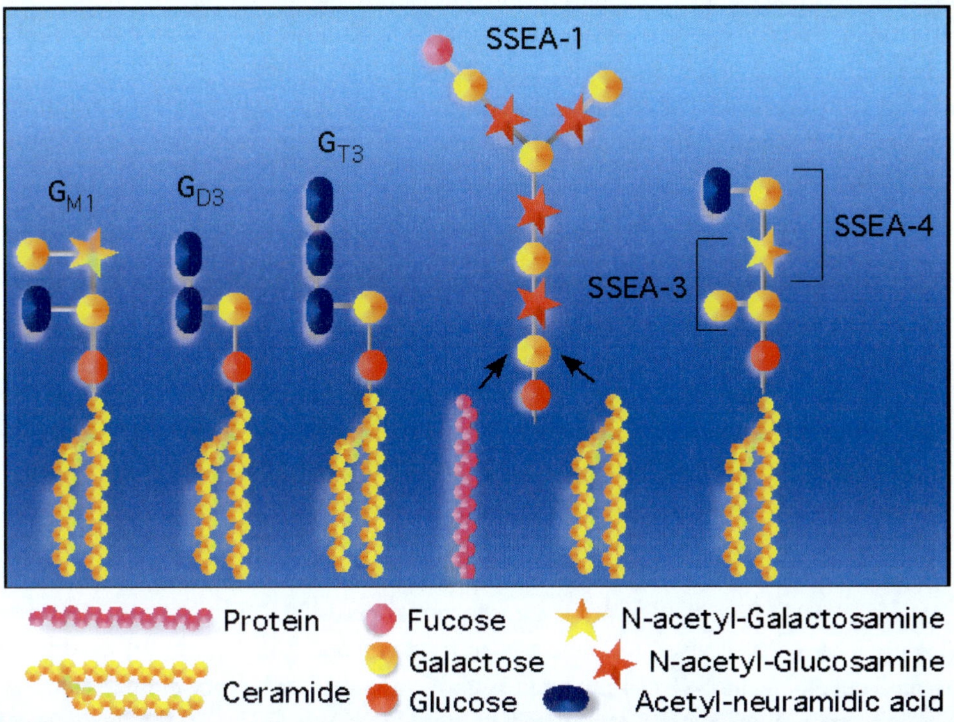

Figure 2. Schematic representation of the glycolipid structures which expression has been identified on neural stem/progenitor cells in human and rodent (adapted from [11, 12]).

Since the majority of neuroepithelial cells express the neural stem and progenitor cell markers nestin and Sox1, cell isolation based on expression of SSEA-1 revealed heterogeneity in the cells within the early neuroepithelium; SSEA-1^{+} cells were highly proliferative whereas the SSEA-1^{-} cells were not and they developed into neuron upon differentiation. This demonstrated that the SSEA-1+ subset contained a population of neural stem cells, whereas the SSEA-1^{-} cells comprised of committed neuronal cells [17,23]. Later during development, SSEA-1 expression becomes restricted to highly neurogenic brain areas (the cerebellum, the hippocampus, and the basal forebrain) in addition to germinative areas of the developing spinal cord [17, 24]. As the nervous system matures (from E14 till adulthood), SSEA-1 exhibits higher levels of expression in proliferative areas of the brain: the ventricular zone (VZ) which will develop into the subventricular zone (SVZ) in the adult, and the hippocampal dentate gyrus. In addition, it is also identified in some cells in the white matter tracts, the rostral migratory stream, the corpus callosum, and the cerebellum [15,16,18,22,25]. At E14, virtually all the SSEA-1^{+} cells express nestin and the glial fibrillary acidic protein, GFAP [17, 26-28].

In the adult mouse brain, SSEA-1 is expressed by a small subset of cells within the SVZ that express GFAP, a marker for adult neural stem cells [16, 26, 29]. Therefore, SSEA-1 allows one to distinguish between the GFAP-expressing adult neural stem cells and the non-neurogenic GFAP-expressing astrocytes [26]. However, SSEA-1/Le^{x} expression is not strictly limited to a subset of neural stem cells, given it was also identified in a subset of astrocytes, tanycytes and in a small subset of neurons [16, 30-32].

In human, SSEA-1 expression has only been reported in the developing brain so far [33]. Its expression in the adult, however, has not yet been documented. In human fetuses, SSEA-1-immunoreactivity was localized in the lateral geniculate nucleus and at low level in the VZ of the dorso-lateral wall of the telencephalon but it was not identified in the cerebral cortex at neonatal stage [33, 34].

1.2. The Stage Specific Embryonic Antigen 3 and 4

Further biochemistry analysis revealed that both SSEA-3 and SSEA-4 antibodies recognized two distinct antigenic sequences on a unique globo-series epitope (see schematic representation in figure 2) expressed on embryonic carcinoma cells and ES cells in both mouse and human [11,35,36]. Like SSEA-1, SSEA-4 represents a reliable marker to monitor the differentiation of mouse and human ES cells. It is important to note, however that the expression profiles of SSEA-1 *versus* SSEA-4 differ between species. SSEA-4 is expressed by undifferentiated human ES cells but not by undifferentiated mouse ES cells, which in contrast, express SSEA-1 at this developmental stage. As differentiation proceeds, human ES cells lose the expression of SSEA-4 [35, 37-39]. SSEA-4 has been identified in human developing CNS, in both the forebrain and the spinal cord [33,40]. It has also been used to label neurosphere cells derived from the human forebrain and spinal cord [40]. During human embryonic development, SSEA-4 immunoreactivity was detected in proliferative areas of the developing brain, but is rarely co-detected with SSEA-1 [33]. At the sixth week of development, strong immunoreactivity for SSEA-4 was identified in the midline and the medial regions of the forebrain. During the seventh week, SSEA-4 expression was detected in restricted areas of the rhombencephalon, the mesencephalon, the diencephalon and the telencephalic wall. By the end of week 7, SSEA-4 is localized in the VZ and the SVZ in the telencephalon. Finally, at week 9, no evident SSEA-4 staining appears in the VZ of the pallium whereas a strong SSEA-4 staining is present in the cortical plate/marginal zone region [33]. A small subset of immature neural cells in the third-trimester human spinal cord were found to be SSEA-4^{+} [40]. Like for SSEA-1, the expression of SSEA-4 is not only restricted to a subset of proliferative neural cells, SSEA-4 has also been described in the dorsal root ganglia (DRG) [41] and in the dorsal aspects of the spinal cord [42]. In the DRG, SSEA-4 labels a subpopulation of neuronal cells whereas in the spinal cord SSEA-4 labels primary afferent sensorial projection that is likely to be the large-diameter cutaneous mechanoreceptors [42, 43]. To the author's knowledge, the expression of SSEA-4 in rodent forebrain has not been yet described.

1.3. Functions of the Stage Specific Embryonic Antigens

Several functions have been attributed to SSEA-1 whereas functions of SSEA-4 in the developing CNS still remain unknown. Such functions appear to vary according to the developmental stage, its level of expression, and possibly the nature of the SSEA-1 carrier that can be either glycolipids or glycoproteins [15-17,44,45]. In ES cells, SSEA-1 was proposed to play a role during compaction, a flattening of the cell surface of the round 8-cell pre-implantation embryo that as a result would increase the cell surface contacts of the cells [46,47]. In early developmental phases of the nervous system, SSEA-1 expression on the cell surface of proliferative neuroepithelial cells, may bind to growth factors like the fibroblast growth factor (FGF), a potent mitogen for neural stem cells [48, 49]. In the embryonic brain, SSEA-1 on its form associated to a protein carrier has been shown to bind to Wnt-1 [17]. One of these protein carriers has recently been identified and corresponded to β1-integrin, a molecule involved in neural stem and progenitor cell adhesion [50]. In the adult SVZ, SSEA-1 is likely to be associated with a glycolipid carrier [16]. However, free SSEA-1 is able to bind the basic FGF or bFGF [48,49] and possibly stimulates mitogenic effects of bFGF on ES cells, as well as neural stem and progenitor cells [49,51]. Given that SSEA-1 is secreted and associated with the extra cellular matrix, it may be an important factor in the neural stem cell niche, binding factors such as Wnt-1 that are important for stem cell growth.

Carbohydrate epitopes may play different roles depending on whether they are covalently linked to lipid or protein core structures. On lipid carrier molecules they are believed to play direct roles in cell adhesion [52]. In the neuroepithelium, SSEA-1 can be either bound to carbohydrate or be soluble *i.e.* released in the extra cellular matrix [17]. A role for carbohydrates in cell proliferation within the CNS has been suggested by Santos-Benito et al [53] who found that treatment of astrocytes, gliomas, and neuroblastomas with fucosyllactose trisaccharides (structurally related to Le^{x}) inhibits mitosis.

In Pax6 mutant rat, Le^{x} epitope expression by cortical cells is lost (due to expression and therefore activity of α-fucosyltransferase IX that is dramatically decreased in the rat cortical cells) [54]. Moreover in this mutant, cortical cell adhesion is lost and neural cell migration from the striatum into the cortex is strongly enhanced [55, 56]. This suggests that Pax6 may be involved in the regulation of Le^{x} epitope expression in the cortical cells of the developing rodent.

2. Gangliosides

As described above, gangliosides are a large family of GSLs containing sialyl acid residues that are abundant in the vertebrate nervous system [14]. Here we will list and detail a few of the gangliosides that have been investigated in neural stem and progenitor cells.

2.1. Monosialogangliosides

G_{M1} antigen (see schematic representation in figure 2), which is specifically recognized by cholera toxin B subunit, has been identified in neural stem and progenitor cells within the embryonic mouse forebrain but not in the spinal cord [27,57]. Flow cytometry analysis revealed that even if cholera toxin B binds G_{M1} with high affinity [58], expression level of this ganglioside (estimated by fluorescence intensity) was still lower than SSEA-1 on the cell surface of mouse embryonic neural precursor cells [57]. Another form of monosialoganglioside, G_{M3}, has been shown to mark differentiating oligodendrocyte-type 2 astrocyte progenitors – also referred to O2A progenitors – in addition to mature oligodendrocytes [59, 60].

2.2. Disialogangliosides

Twenty years ago, a cell surface antigen – named D1.1 antigen – was identified as a ganglioside associated with an O-acetylated sialic acid residue, an O-acetylated G_{D3} ganglioside [61]. Given its expression early in the neural development of the rat (in the early neuroepithelium, within the neural plate and the neural tube), D1.1 antigen was proposed as a marker for undifferentiated neural cells. Further histological analysis demonstrated that D1.1 antigen was identified in both developing and adult rat nervous systems [62,63]. At E12 in the rat, most the telencephalon was immnuoreactive for anti-D1.1 whereas at E18 immunoreactivity was only identified in the VZ but not in the mantle and marginal zones of the telencephalon [62]. In addition, antigen D1.1 was also observed in the developing cerebellum in rat; like the telencephalon, D1.1 expression was localized in the germinative zone, in the external granule layer known to comprise most proliferating cells [62]. As the cerebellum matures, expression of D1.1 decreases and disappears with the external granule layer by post-natal day 20 [62]. However, like SSEA-4, D1.1 is not exactly restricted to neural precursor cells, given it was also visualized on differentiated sensory neurons in the DRG. At E8-E9 in rat, D1.1 is expressed in the anterior segment of the neural tube just prior to and after its closure at E10-E11. Later during development till adulthood, expression of D1.1 was found restricted to SVZ cells [63]. Based on the fact that the D1.1 antibody does not bind the most common gangliosides (G_{T1}, G_{D1a}, GD_{1b}, G_{D2}, G_{D3}, G_{M1}, G_{M2}, and G_{M3}), the authors implied that the ganglioside recognised by anti-D1.1 might only be expressed transiently during neurogenesis [62].

Parallel to D1.1 antigen identification, another ganglioside – G_{D3} ganglioside (CD60a) – (see schematic representation in figure 2) is highly expressed in the SVZ, in the granular cell layer of the hippocampus, and below the pial surface of the cerebral cortex of post-natal rat. More recently, it was suggested that G_{D3} represents the most abundant gangliosides in E12-E14 mouse brains whereas its expression drastically decreased at E16 [64]. In the postnatal mouse, G_{D3}-expressing cells were identified in $GFAP^{+}$ radial glial cells in the cerebral cortex, while in the adult it labelled astrocytes present in the cortex, corpus callosum and neural stem cells within the SVZ [65,66]. In the adult rat, a small subset of SVZ cells was found G_{D3}^+: these cells were located just underneath the ependymal layer, an area known to comprise neural stem cells [16,29,67]. From cell cultures established from neonatal mouse cerebral hemispheres, G_{D3} marks the oligodendrocyte progenitor cells [68]. However, G_{D3} is not completely restricted to proliferating cells given that it represents one of the major ganglioside constituents of human oligodendrocytes [69].

2.3. Trisialogangliosides

In 1979, a monoclonal antibody (A2B5 antibody) raised to plasma membrane of chicken embryonic retinal cells was developed, and subsequent biochemical analysis revealed that A2B5 antibody recognized trisialogangliosides G_{T3} and o-acetylG_{T3} whose schematic representations are shown in figure 2 [70,71]. On rodent tissue sections and cell cultures established from CNS, A2B5 labelled immature astrocytes and oligodendrocytes, a population of bipotential glial restricted precursor cells [71-74]. Given their potential to give rise to either oligodendrocytes or type 2 astrocytes depending on culture conditions, these glial restricted precursor cells were named O2A progenitor cells [71, 72]. A2B5 was further used *in vitro* to enrich for oligodendrocyte and astrocyte precursor cells, by immunopanning. This method consists of plating a cell suspension onto a culture dish coated with the anti-A2B5 so that only A2B5$^+$ cells remained attached to the culture dish [74].

Expression of A2B5 has been investigated in embryonic and adult rodents [64, 73, 75]. In the embryonic rat (at E15), two populations of A2B5$^+$ cells were identified in the ventral and in the dorsal aspects of the telencephalon [75]. In the ventral telencephalon only, A2B5$^+$ cells coexpressed Olig2, a specific marker for oligodendrocyte progenitor cells. Flow cytometry analysis of these rat telencephalic cells with anti-A2B5 and antibody against the polysialylated form of Neural Cell Adhesion Molecule (PSA-NCAM), revealed that a large fraction of these A2B5$^+$ cells also co-labelled with anti-PSA-NCAM demonstrating heterogeneity in the A2B5$^+$ cell population [75]. The A2B5$^+$/PSA-NCAM$^+$ defined a subset of neuronal progenitor cells, the A2B5$^+$/PSA-NCAM$^-$ cell subset gave rise to both astrocytes and oligodendrocytes upon differentiation, demonstrating that A2B5 in association with PSA-NCAM antibody identify either neuronal of glial restricted precursor cell subsets [75]. In the adult, A2B5 marks a heterogeneous population of proliferating oligodendrocyte precursor cells in the forebrain white matter: this subpopulation was described as heterogeneous based on receptor expression for mitogens like the platelet growth factor receptor (PDGF-αR), insulin growth factor receptor (IGF-1R) and fibroblast growth factor receptor (FGF-1R) and also due to their response to PDGF, IGF and bFGF [73].

2.4. Functions of the Gangliosides

Cellular functions of the gangliosides still remain unknown. Due to their abundance in lipid rafts - a lipid microenvironment at the surface of the cells - gangliosides are likely to play important functions in cell signalling and possibly cell adhesion [50, 76]. In particular, G_{D3} gangliosides have been shown to affect proliferation of a cell line of neural precursor cells derived from neonatal mouse cerebellar cortex, by decreasing mitogenic effects of the epidermal growth factor or EGF [77]. More recently, Yanagisawa and collaborators [50] treated neurosphere cells derived from mouse embryonic telencephalon with an inhibitor of glucosylceramide synthesis: such a treatment lead to the elimination of the G_{D3} expression of neural precursor cells and to a decrease in their proliferation rate in response to the mitogens bFGF and leukaemia inhibitory factor or LIF. This reduction in the proliferation rate of the neural precursor cells was due to a repression of the Ras-MAPK pathway, a cytoplasmic signal transducer that is activated following binding of the mitogens (either bFGF or LIF) to their receptors leading to activation of the cascade of events conducting to cell division. Given that Ras-MAPK are particularly abundant in lipid rafts, it is possible that gangliosides present in the lipid rafts may act as regulator of the neural stem and progenitor cell division [78, 79].

Glycolipids: New Insights in Neural Stem Cell Field

In the last decade, cell replacement therapy based on delivering specialized (e.g. oligodendrocyte precursor cells) or unspecialized cells (neural stem cells) to replace missing specialized cells that are at the origin of the disease, was proposed as a strategy to cure several neurological disorders like amyotrophic lateral sclerosis or ALS, Parkinson's disease, and Huntingdon's disease [80-83]. Neural stem and progenitor cells persist in the adult, in both normal and pathologic conditions [29, 84, 85]. Endogenous neural stem and progenitor cells constantly renew specific neuronal or glial cell subsets throughout life, however they failed to self-repair the injured nervous system. For example, in multiple sclerosis or following spinal cord injury, endogenous oligodendrocyte precursor cells were identified in the injured nervous system areas but did not differentiate or repair by promoting axonal outgrowth and via the formation of new myelin sheath [86-91]. To overcome the limitations of endogenous neural stem/progenitor cells to self-repair, cell-based therapy could represent one of the strategies to promote nervous system repair. Therefore, identification of cell surface markers specific for a given cell lineage or for subsets of neural stem cells appears necessary in order to select, isolate using flow cytometry and enrich for a specific cell type prior to transplantation. Using "purified" populations of stem or progenitor cells will allow investigators to obtain a relatively homogeneous subset of cells and as a result will probably increase the chance of reproducing therapeutic effects. Finally, neural stem and progenitor cells are expandable *in vitro* and therefore limit the need to harvest tissue. As a result, cell surface markers should allow identification of a relatively homogeneous population of cells (*e.g.* neural stem cells or oligodendrocyte progenitor cells) from both tissue (*in vivo*) and/or after expansion *in vitro* for their prospective isolation. Critically, these cells should not be immunogenic and therefore induce immune rejection against the grafted cells, and they should migrate towards the injured area, integrate and repair the damage.

1. Cell Surface Marker and Stem Cell or Progenitor Cell Selection

1.1. Neural Stem Cell Selection: The SSEAs

Given their expression on the external surface of the cells, SSEA-1 and SSEA-4 represents very valuable markers that can be used to enrich for homogenous subset of neural stem cells. Although SSEA-1 has been used for the positive selection and enrichment of neural stem cells in both embryonic and adult mouse and in fetal human brains, expression of SSEA-4 on human fetal neural stem cells has only been reported recently [16,17,33,40]. A functional *in vitro* assay to demonstrate self-renewal and multipotency – the hallmark properties of neural stem cells – is the neurosphere-forming assay [92-94]. This method consists of plating single cells at a low density in order to avoid cell reaggregation, in a medium supplemented in mitogens, and to follow capacity of individual cell to divide *i.e.* to generate a clone or primary neurosphere. To evaluate their self-renewing potential, each clone is dissociated to a single cell, and each individual cell is inspected for its capacity to form new clones or secondary neurospheres [95]. To test whether SSEA-1-expressing cells from both the developing and adult CNS can produce clones, SSEA-1$^+$ cells were isolated by fluorescence activated cell sorting or FACS and plated in mitogen supplemented medium. In mouse embryo and adult, the subset of cells expressing SSEA-1 was enriched in multipotent sphere-forming cells compared to the cells that did not express SSEA-1 [16, 17, 96]. In human fetal brain, both SSEA-1 and SSEA-4 were found enriched in a large population of cells expressing CD133, (a glycoprotein also referred to as Prominin in mice), one of the most common cell surface marker used to isolate by FACS or immunomagnetic separation a subset of cells comprising most of the sphere-forming cells [28,33,96-101]. In addition to their enrichment in the CD133$^+$ cell fraction, SSEA-1 and SSEA-4 appeared to label two distinct populations of sphere-forming cells [33]. In contrast, cells expressing CD133 but negative for either SSEA-1 or SSEA-4 comprised of very few sphere-forming cells demonstrating that a combination of CD133 and SSEA-1 or CD133 and SSEA-4 allowed for further enriching for distinct subsets of sphere-forming cells. Although SSEA-1 and SSEA-4-expressing cell subsets labelled most self-renewal cells, multipotency has only been reported for SSEA-4- but not SSEA-1-expressing cells [33].

Neurosphere assay has been widely used to expand under EGF and/or bFGF stimulation neural stem cells from either embryonic or adult brains [92, 93, 102-104]. Neurosphere cultures established from the embryonic mouse and human sustain the expansion of a small subset of cells expressing SSEA-1 or SSEA-4, respectively, that are enriched in sphere-forming cells [17, 28, 33, 96].

Therefore SSEA-1 and SSEA-4 represent valuable markers to be used for the enrichment of multipotent and self-renewal cells, and for their prospective used in cell-based therapy.

1.2. Oligodendrocyte Progenitor Cells: A2B5

The cell surface epitope A2B5 has been used in several studies for the positive selection of glial restricted progenitor cells, a cell subset with potential to give rise to both astrocytes and oligodendrocytes from both embryonic rat and adult human [75,105]. In a neurosphere-forming assay, the A2B5$^+$/PSA-NCAM$^-$ cells isolated by flow cytometry from embryonic rat telencephalon gave rise to clones that upon differentiation only generated glial cells (astrocytes and oligodendrocytes) but no neurons, demonstrating that A2B5$^+$ combined with PSA-NCAM allows for the enrichment of glial restricted progenitor cells [75]. Moreover, A2B5 not only defined a subset of glial restricted progenitor cells *in vivo*, but also *in vitro* after expansion of the rat telencephalic neural stem/progenitor cells in a defined (serum-free) medium [75].

2. Cell Surface Antigen and Immunogenicity

Because of their presentation on the cell surface of stem/progenitor cells, glycolipids may represent antigenic components susceptible to be immunogenic. For example, Le^x is expressed on the cell surface of erythrocytes and defined antigenic determinants of blood groups. Therefore it appears that using such glycolipids as cell surface epitopes for the positive selection of subsets of stem and/or progenitor cell for cell-based therapy is perhaps not without risks. Corti et al [96] isolated by flow cytometry a subset of neurosphere cells derived from the mouse adult SVZ, based on SSEA-1 (Le^x) expression. Purified neural stem cells were then grafted into the spinal cord of a transgenic mouse line over-expressing the human form of the superoxide dismutase (SOD1) reproducing some of the symptoms occurring in ALS (*i.e.* progressive degeneration of motor neurons in the brain and spinal cord conducting to progressive paralysis). Grafted SSEA-1+ neural stem/progenitor cells survived at least a month after transplantation and did not produce any sign of immune response around the graft (inspected in this study by the absence of lymphocyte T CD4 or CD8, or natural killer cells in areas containing host cells). Similar results were obtained from transplanted subsets of glial restricted progenitor cells of rat origin (based on A2B5 expression) into animal model of demyelination [75] demonstrating the feasibility of using cell subsets expressing glycolipid cell surface antigens for cell-based therapy.

3. SSEA-1 and A2B5: Biomarkers for Therapeutic Cells?

In addition to survival and integration into experimental models of injured CNS, A2B5-expressing and SSEA-1-expressing cells have demonstrated potential to repair the damage [75,96,105]. Purified A2B5-expressing cells were transplanted into demyelinated lesions in the adult rat brain or into the forebrain of genetically modified postnatal mouse strain carrying a mutation in the major myelin protein, the myelin basic protein or MBP [75,105]. Knockout mice for MBP named *shiverer* mice are characterized by a decreased level of CNS myelination and a lack of compacted myelin sheath. Transplanted A2B5$^+$ cells were able to develop into functional oligodendrocytes and even to form compacted myelin sheath around axonal bundles of the recipient [75,105]. Transplantation of undifferentiated cells expressing SSEA-1 into animal model of ALS gave rise to neuronal cells including some motor neurons and even promoted neuroprotection by secreting factors such as vascular endothelial growth factor or VEGF and IGF [96].

Conclusions

One of the major limitations in studying neural stem cell biology has been the lack of specific markers for their prospective identification and isolation for further investigation in *in vitro* or *in vivo* assays. Because of their potential to self-renew giving rise to a large number of identical cells and to generate either neuronal or glial cells, neural stem cells represents a very attractive source of cells for alternative cell-based therapy of several neurological disorders or to repair CNS injury. Glycolipids, which are exposed on the external leaflet of the cell membrane and participate in several biological functions like adhesion and migration, have been widely used as biomarkers to monitor embryonic stem cell differentiation in mammals. Their recent identification on the cell surface of stem or progenitor cells in both developing and adult brains, offers new perspectives in the neural stem cell field and for their potential use in cell-based therapy.

References

[1] Kleene, R. and Schachner, M. (2004). Glycans and neural cell interactions. *Nat. Rev. Neurosci.* 5, 195-208.

[2] Karlsson, K. A. (1970). On the chemistry and occurrence of sphingolipid long-chain bases. *Chem. Phys. Lipids* 5, 6-43.

[3] Steinman, R. M., Mellman, I. S., Muller, W. A., and Cohn, Z. A. (1983). Endocytosis and the recycling of plasma membrane. *J. Cell Biol.* 96, 1-27.

[4] Tettamanti, G. (2004). Ganglioside/glycosphingolipid turnover: new concepts. *Glycoconj. J.* 20, 301-317.

[5] Csiza, C. K. (1982). Lipid class analysis of the central nervous system of myelin-deficient Wistar rats. *J. Lipid Res.* 23, 720-725.

[6] Artzt, K., Dubois, P., Bennett, D., Condamine, H., Babinet, C., and Jacob, F. (1973). Surface antigens common to mouse cleavage embryos and primitive teratocarcinoma cells in culture. *Proc. Natl. Acad. Sci. USA* 70, 2988-2992.

[7] Dewey, M. J., Gearhart, J. D., and Mintz, B. (1977). Cell surface antigens of totipotent mouse teratocarcinoma cells grown in vivo: their relation to embryo, adult, and tumor antigens. *Dev. Biol.* 55, 359-374.

[8] Gooding, L. R. and Edidin, M. (1974). Cell surface antigens of a mouse testicular teratoma. Identification of an antigen physically associated with H-2 antigens on tumor cells. *J. Exp. Med.* 140, 61-78.

[9] Stern, P. L., Willison, K. R., Lennox, E., Galfre, G., Milstein, C., Secher, D., and Ziegler, A. (1978). Monoclonal antibodies as probes for differentiation and tumor-associated antigens: a Forssman specificity on teratocarcinoma stem cells. *Cell* 14, 775-783.

[10] Solter, D. and Knowles, B. B. (1978). Monoclonal antibody defining a stage-specific mouse embryonic antigen (SSEA-1). *Proc. Natl. Acad. Sci. USA* 75, 5565-5569.

[11] Kannagi, R., Cochran, N. A., Ishigami, F., Hakomori, S., Andrews, P. W., Knowles, B. B., and Solter, D. (1983). Stage-specific embryonic antigens (SSEA-3 and -4) are

epitopes of a unique globo-series ganglioside isolated from human teratocarcinoma cells. *EMBO J.* 2, 2355-2361.

[12] Nudelman, E., Hakomori, S., Knowles, B. B., Solter, D., Nowinski, R. C., Tam, M. R., and Young, W. W., Jr. (1980). Monoclonal antibody directed to the stage-specific embryonic antigen (SSEA-1) reacts with branched glycosphingolipid similar in structure to Ii antigen. *Biochem. Biophys. Res. Commun.* 97, 443-451.

[13] Gooi, H. C., Feizi, T., Kapadia, A., Knowles, B. B., Solter, D., and Evans, M. J. (1981). Stage-specific embryonic antigen involves alpha 1 goes to 3 fucosylated type 2 blood group chains. *Nature* 292, 156-158.

[14] Hakomori, S. (1990). Bifunctional role of glycosphingolipids. Modulators for transmembrane signaling and mediators for cellular interactions. *J. Biol. Chem.* 265, 18713-18716.

[15] Allendoerfer, K. L., Magnani, J. L., and Patterson, P. H. (1995). FORSE-1, an antibody that labels regionally restricted subpopulations of progenitor cells in the embryonic central nervous system, recognizes the Le(x) carbohydrate on a proteoglycan and two glycolipid antigens. *Mol. Cell Neurosci.* 6, 381-395.

[16] Capela, A. and Temple, S. (2002). LeX/ssea-1 is expressed by adult mouse CNS stem cells, identifying them as nonependymal. *Neuron* 35, 865-875.

[17] Capela, A. and Temple, S. (2006). LeX is expressed by principle progenitor cells in the embryonic nervous system, is secreted into their environment and binds Wnt-1. *Dev. Biol.* 291, 300-313.

[18] Fox, N., Damjanov, I., Martinez-Hernandez, A., Knowles, B. B., and Solter, D. (1981). Immunohistochemical localization of the early embryonic antigen (SSEA-1) in postimplantation mouse embryos and fetal and adult tissues. *Dev. Biol.* 83, 391-398.

[19] Mai, J. K., Andressen, C., and Ashwell, K. W. (1998). Demarcation of prosencephalic regions by CD15-positive radial glia. *Eur. J. Neurosci.* 10, 746-751.

[20] Solter, D. and Knowles, B. B. (1979). Developmental stage-specific antigens during mouse embryogenesis. *Curr. Top. Dev. Biol.* 13 Pt 1, 139-165.

[21] Temple, S. (2001). The development of neural stem cells. *Nature* 414, 112-117.

[22] Yamamoto, M., Boyer, A. M., and Schwarting, G. A. (1985). Fucose-containing glycolipids are stage- and region-specific antigens in developing embryonic brain of rodents. *Proc. Natl. Acad. Sci. U.S.A* 82, 3045-3049.

[23] Lendahl, U., Zimmerman, L. B., and McKay, R. D. (1990). CNS stem cells express a new class of intermediate filament protein. *Cell* 60, 585-595.

[24] Dodd, J. and Jessell, T. M. (1986). Cell surface glycoconjugates and carbohydrate-binding proteins: possible recognition signals in sensory neurone development. *J. Exp. Biol.* 124, 225-238.

[25] Tole, S. and Patterson, P. H. (1995). Regionalization of the developing forebrain: a comparison of FORSE-1, Dlx-2, and BF-1. *J. Neurosci.* 15, 970-980.

[26] Imura, T., Nakano, I., Kornblum, H. I., and Sofroniew, M. V. (2006). Phenotypic and functional heterogeneity of GFAP-expressing cells in vitro: differential expression of LeX/CD15 by GFAP-expressing multipotent neural stem cells and non-neurogenic astrocytes. *Glia* 53, 277-293.

[27] Liour, S. S., Dinkins, M. B., Su, C. Y., and Yu, R. K. (2005). Spatiotemporal expression of GM1 in murine medial pallial neural progenitor cells. *J. Comp. Neurol.* 491, 330-338.

[28] Panchision, D. M., Chen, H. L., Pistollato, F., Papini, D., Ni, H. T., and Hawley, T. S. (2007). Optimized flow cytometric analysis of central nervous system tissue reveals novel functional relationships among cells expressing CD133, CD15, and CD24. *Stem Cells* 25, 1560-1570.

[29] Doetsch, F., Caille, I., Lim, D. A., Garcia-Verdugo, J. M., and Alvarez-Buylla, A. (1999). Subventricular zone astrocytes are neural stem cells in the adult mammalian brain. *Cell* 97, 703-716.

[30] Ashwell, K. W. and Mai, J. K. (1997). Developmental expression of the CD15-epitope in the brainstem and spinal cord of the mouse. *Anat. Embryol. (Berl)* 196, 13-25.

[31] Bartsch, D. and Mai, J. K. (1991). Distribution of the 3-fucosyl-N-acetyl-lactosamine (FAL) epitope in the adult mouse brain. *Cell Tissue Res.* 263, 353-366.

[32] Gocht, A., Struckhoff, G., and Lhler, J. (1996). CD15-containing glycoconjugates in the central nervous system. *Histol. Histopathol.* 11, 1007-1028.

[33] Barraud, P., Stott, S., Mollgard, K., Parmar, M., and Bjorklund, A. (2007). In vitro characterization of a human neural progenitor cell coexpressing SSEA4 and CD133. *J. Neurosci. Res.* 85, 250-259.

[34] Schwartz, P. H., Bryant, P. J., Fuja, T. J., Su, H., O'Dowd, D. K., and Klassen, H. (2003). Isolation and characterization of neural progenitor cells from post-mortem human cortex. *J. Neurosci. Res.* 74, 838-851.

[35] Ginis, I., Luo, Y., Miura, T., Thies, S., Brandenberger, R., Gerecht-Nir, S., Amit, M., Hoke, A., Carpenter, M. K., Itskovitz-Eldor, J., and Rao, M. S. (2004). Differences between human and mouse embryonic stem cells. *Dev. Biol.* 269, 360-380.

[36] Krupnick, J. G., Damjanov, I., Damjanov, A., Zhu, Z. M., and Fenderson, B. A. (1994). Globo-series carbohydrate antigens are expressed in different forms on human and murine teratocarcinoma-derived cells. *Int. J. Cancer* 59, 692-698.

[37] Carpenter, M. K., Rosler, E., and Rao, M. S. (2003). Characterization and differentiation of human embryonic stem cells. *Cloning Stem Cells* 5, 79-88.

[38] Henderson, J. K., Draper, J. S., Baillie, H. S., Fishel, S., Thomson, J. A., Moore, H., and Andrews, P. W. (2002). Preimplantation human embryos and embryonic stem cells show comparable expression of stage-specific embryonic antigens. *Stem Cells* 20, 329-337.

[39] Schulz, T. C., Palmarini, G. M., Noggle, S. A., Weiler, D. A., Mitalipova, M. M., and Condie, B. G. (2003). Directed neuronal differentiation of human embryonic stem cells. *BMC. Neurosci.* 4, 27.

[40] Piao, J. H., Odeberg, J., Samuelsson, E. B., Kjaeldgaard, A., Falci, S., Seiger, A., Sundstrom, E., and Akesson, E. (2006). Cellular composition of long-term human spinal cord- and forebrain-derived neurosphere cultures. *J. Neurosci. Res.* 84, 471-482.

[41] Holford, L. C., Case, P., and Lawson, S. N. (1994). Substance P, neurofilament, peripherin and SSEA4 immunocytochemistry of human dorsal root ganglion neurons obtained from post-mortem tissue: a quantitative morphometric analysis. *J. Neurocytol.* 23, 577-589.

[42] Alvarez, F. J., Villalba, R. M., Zerda, R., and Schneider, S. P. (2004). Vesicular glutamate transporters in the spinal cord, with special reference to sensory primary afferent synapses. *J. Comp. Neurol.* 472, 257-280.

[43] Jessell, T. M. and Dodd, J. (1985). Structure and expression of differentiation antigens on functional subclasses of primary sensory neurons. *Philos. Trans. R. Soc. Lond B Biol. Sci.* 308, 271-281.

[44] Allendoerfer, K. L., Durairaj, A., Matthews, G. A., and Patterson, P. H. (1999). Morphological domains of Lewis-X/FORSE-1 immunolabeling in the embryonic neural tube are due to developmental regulation of cell surface carbohydrate expression. *Dev. Biol.* 211, 208-219.

[45] Tole, S., Kaprielian, Z., Ou, S. K., and Patterson, P. H. (1995). FORSE-1: a positionally regulated epitope in the developing rat central nervous system. *J. Neurosci.* 15, 957-969.

[46] Ducibella, T., Ukena, T., Karnovsky, M., and Anderson, E. (1977). Changes in cell surface and cortical cytoplasmic organization during early embryogenesis in the preimplantation mouse embryo. *J. Cell Biol.* 74, 153-167.

[47] Richa, J., Damsky, C. H., Buck, C. A., Knowles, B. B., and Solter, D. (1985). Cell surface glycoproteins mediate compaction, trophoblast attachment, and endoderm formation during early mouse development. *Dev. Biol.* 108, 513-521.

[48] Dvorak, P., Hampl, A., Jirmanova, L., Pacholikova, J., and Kusakabe, M. (1998). Embryoglycan ectodomains regulate biological activity of FGF-2 to embryonic stem cells. *J. Cell Sci.* 111 (Pt 19), 2945-2952.

[49] Jirmanova, L., Pacholikova, J., Krejci, P., Hampl, A., and Dvorak, P. (1999). O-linked carbohydrates are required for FGF-2-mediated proliferation of mouse embryonic cells. *Int. J. Dev. Biol.* 43, 555-562.

[50] Yanagisawa, M., Taga, T., Nakamura, K., Ariga, T., and Yu, R. K. (2005). Characterization of glycoconjugate antigens in mouse embryonic neural precursor cells. *J. Neurochem.* 95, 1311-1320.

[51] Raballo, R., Rhee, J., Lyn-Cook, R., Leckman, J. F., Schwartz, M. L., and Vaccarino, F. M. (2000). Basic fibroblast growth factor (Fgf2) is necessary for cell proliferation and neurogenesis in the developing cerebral cortex. *J. Neurosci.* 20, 5012-5023.

[52] Eggens, I., Fenderson, B. A., Toyokuni, T., and Hakomori, S. (1989). A role of carbohydrate-carbohydrate interaction in the process of specific cell recognition during embryogenesis and organogenesis: a preliminary note. *Biochem. Biophys. Res. Commun.* 158, 913-920.

[53] Santos-Benito, F. F., Fernandez-Mayoralas, A., Martin-Lomas, M., and Nieto-Sampedro, M. (1992). Inhibition of proliferation of normal and transformed neural cells by blood group-related oligosaccharides. *J. Exp. Med.* 176, 915-918.

[54] Shimoda, Y., Tajima, Y., Osanai, T., Katsume, A., Kohara, M., Kudo, T., Narimatsu, H., Takashima, N., Ishii, Y., Nakamura, S., Osumi, N., and Sanai, Y. (2002). Pax6 controls the expression of Lewis x epitope in the embryonic forebrain by regulating alpha 1,3-fucosyltransferase IX expression. *J. Biol. Chem.* 277, 2033-2039.

[55] Chapouton, P., Gartner, A., and Gotz, M. (1999). The role of Pax6 in restricting cell migration between developing cortex and basal ganglia. *Development* 126, 5569-5579.

[56] Stoykova, A., Gotz, M., Gruss, P., and Price, J. (1997). Pax6-dependent regulation of adhesive patterning, R-cadherin expression and boundary formation in developing forebrain. *Development* 124, 3765-3777.

[57] Yanagisawa, M., Ariga, T., and Yu, R. K. (2006). Cholera toxin B subunit binding does not correlate with GM1 expression: a study using mouse embryonic neural precursor cells. *Glycobiology* 16, 19G-22G.

[58] Kuziemko, G. M., Stroh, M., and Stevens, R. C. (1996). Cholera toxin binding affinity and specificity for gangliosides determined by surface plasmon resonance. *Biochemistry* 35, 6375-6384.

[59] Yim, S. H., Yavin, E., Hammer, J. A., and Quarles, R. H. (1991). Exogenous GM3 ganglioside stimulates process formation and glycoprotein release by cultured bovine oligodendrocytes. *J. Neurochem.* 57, 2144-2147.

[60] Yim, S. H., Farrer, R. G., Hammer, J. A., Yavin, E., and Quarles, R. H. (1994). Differentiation of oligodendrocytes cultured from developing rat brain is enhanced by exogenous GM3 ganglioside. *J. Neurosci. Res.* 38, 268-281.

[61] Cheresh, D. A., Varki, A. P., Varki, N. M., Stallcup, W. B., Levine, J., and Reisfeld, R. A. (1984). A monoclonal antibody recognizes an O-acylated sialic acid in a human melanoma-associated ganglioside. *J. Biol. Chem.* 259, 7453-7459.

[62] Levine, J. M., Beasley, L., and Stallcup, W. B. (1984). The D1.1 antigen: a cell surface marker for germinal cells of the central nervous system. *J. Neurosci.* 4, 820-831.

[63] Levine, J. M., Beasley, L., and Stallcup, W. B. (1986). Localization of a neurectoderm-associated cell surface antigen in the developing and adult rat. *Brain Res.* 392, 211-222.

[64] Ngamukote, S., Yanagisawa, M., Ariga, T., Ando, S., and Yu, R. K. (2007). Developmental changes of glycosphingolipids and expression of glycogenes in mouse brains. *J. Neurochem.*

[65] Cammer, W. and Zhang, H. (1996). Ganglioside GD3 in radial glia and astrocytes in situ in brains of young and adult mice. *J. Neurosci. Res.* 46, 18-23.

[66] Cammer, W. and Zhang, H. (1996). A reexamination of ganglioside GD3-immunopositive glial cells in the forebrains of rats and mice less than 1 week of age. *J. Histochem. Cytochem.* 44, 143-149.

[67] Goldman, J. E., Hirano, M., Yu, R. K., and Seyfried, T. N. (1984). GD3 ganglioside is a glycolipid characteristic of immature neuroectodermal cells. *J. Neuroimmunol.* 7, 179-192.

[68] Vitry, S., Avellana-Adalid, V., Hardy, R., Lachapelle, F., and Baron-Van Evercooren, A. (1999). Mouse oligospheres: from pre-progenitors to functional oligodendrocytes. *J. Neurosci. Res.* 58, 735-751.

[69] Yu, R. K. and Iqbal, K. (1979). Sialosylgalactosyl ceramide as a specific marker for human myelin and oligodendroglial perikarya: gangliosides of human myelin, oligodendroglia and neurons. *J. Neurochem.* 32, 293-300.

[70] Eisenbarth, G. S., Walsh, F. S., and Nirenberg, M. (1979). Monoclonal antibody to a plasma membrane antigen of neurons. *Proc. Natl. Acad. Sci. USA* 76, 4913-4917.

[71] Farrer, R. G. and Quarles, R. H. (1999). GT3 and its O-acetylated derivative are the principal A2B5-reactive gangliosides in cultured O2A lineage cells and are down-

regulated along with O-acetyl GD3 during differentiation to oligodendrocytes. *J. Neurosci. Res.* 57, 371-380.

[72] Dubois-Dalcq, M. (1987). Characterization of a slowly proliferative cell along the oligodendrocyte differentiation pathway. *EMBO J.* 6, 2587-2595.

[73] Mason, J. L. and Goldman, J. E. (2002). A2B5+ and O4+ Cycling progenitors in the adult forebrain white matter respond differentially to PDGF-AA, FGF-2, and IGF-1. *Mol. Cell Neurosci.* 20, 30-42.

[74] Mi, H. and Barres, B. A. (1999). Purification and characterization of astrocyte precursor cells in the developing rat optic nerve. *J. Neurosci.* 19, 1049-1061.

[75] Strathmann, F. G., Wang, X., and Mayer-Proschel, M. (2007). Identification of two novel glial-restricted cell populations in the embryonic telencephalon arising from unique origins. *BMC. Dev. Biol.* 7, 33.

[76] Simons, K. and Toomre, D. (2000). Lipid rafts and signal transduction. *Nat. Rev. Mol. Cell Biol.* 1, 31-39.

[77] Yanagisawa, M., Liour, S. S., and Yu, R. K. (2004). Involvement of gangliosides in proliferation of immortalized neural progenitor cells. *J. Neurochem.* 91, 804-812.

[78] Mineo, C., James, G. L., Smart, E. J., and Anderson, R. G. (1996). Localization of epidermal growth factor-stimulated Ras/Raf-1 interaction to caveolae membrane. *J. Biol. Chem.* 271, 11930-11935.

[79] Wu, C., Butz, S., Ying, Y., and Anderson, R. G. (1997). Tyrosine kinase receptors concentrated in caveolae-like domains from neuronal plasma membrane. *J. Biol. Chem.* 272, 3554-3559.

[80] Bjorklund, A. and Lindvall, O. (2000). Cell replacement therapies for central nervous system disorders. *Nat. Neurosci.* 3, 537-544.

[81] Franklin, R. J. and Blakemore, W. F. (1997). Transplanting oligodendrocyte progenitors into the adult CNS. *J. Anat.* 190 (Pt 1), 23-33.

[82] Keirstead, H. S. (2005). Stem cells for the treatment of myelin loss. *Trends Neurosci.* 28, 677-683.

[83] Lindvall, O., Kokaia, Z., and Martinez-Serrano, A. (2004). Stem cell therapy for human neurodegenerative disorders-how to make it work. *Nat. Med.* 10 Suppl, S42-S50.

[84] Franklin, R. J. and Blakemore, W. F. (1997). To what extent is oligodendrocyte progenitor migration a limiting factor in the remyelination of multiple sclerosis lesions? *Mult. Scler.* 3, 84-87.

[85] Lois, C. and Alvarez-Buylla, A. (1993). Proliferating subventricular zone cells in the adult mammalian forebrain can differentiate into neurons and glia. *Proc. Natl. Acad. Sci. USA* 90, 2074-2077.

[86] Dubois-Dalcq, M., Ffrench-Constant, C., and Franklin, R. J. (2005). Enhancing central nervous system remyelination in multiple sclerosis. *Neuron* 48, 9-12.

[87] Prineas, J. W., Kwon, E. E., Goldenberg, P. Z., Ilyas, A. A., Quarles, R. H., Benjamins, J. A., and Sprinkle, T. J. (1989). Multiple sclerosis. Oligodendrocyte proliferation and differentiation in fresh lesions. *Lab. Invest* 61, 489-503.

[88] Scolding, N. J., Morgan, B. P., and Compston, D. A. (1998). The expression of complement regulatory proteins by adult human oligodendrocytes. *J.Neuroimmunol.* 84, 69-75.

[89] Sim, F. J., Zhao, C., Penderis, J., and Franklin, R. J. (2002). The age-related decrease in CNS remyelination efficiency is attributable to an impairment of both oligodendrocyte progenitor recruitment and differentiation. *J. Neurosci.* 22, 2451-2459.

[90] Suyama, K., Watanabe, M., Sakai, D., Osada, T., Imai, M., and Mochida, J. (2007). Nkx2.2 expression in differentiation of oligodendrocyte precursor cells and inhibitory factors for differentiation of oligodendrocytes after traumatic spinal cord injury. *J. Neurotrauma* 24, 1013-1025.

[91] Zhao, C., Fancy, S. P., Magy, L., Urwin, J. E., and Franklin, R. J. (2005). Stem cells, progenitors and myelin repair. *J. Anat.* 207, 251-258.

[92] Reynolds, B. A. and Weiss, S. (1992). Generation of neurons and astrocytes from isolated cells of the adult mammalian central nervous system. *Science* 255, 1707-1710.

[93] Richards, L. J., Kilpatrick, T. J., Bartlett, P. F., and Murphy, M. (1992). Leukemia inhibitory factor promotes the neuronal development of spinal cord precursors from the neural tube. *J. Neurosci. Res.* 33, 476-484.

[94] Seaberg, R. M. and van der, K. D. (2003). Stem and progenitor cells: the premature desertion of rigorous definitions. *Trends Neurosci.* 26, 125-131.

[95] Barraud, P., Thompson, L., Kirik, D., Bjorklund, A., and Parmar, M. (2005). Isolation and characterization of neural precursor cells from the Sox1-GFP reporter mouse. *Eur. J. Neurosci.* 22, 1555-1569.

[96] Corti, S., Locatelli, F., Papadimitriou, D., Del Bo, R., Nizzardo, M., Nardini, M., Donadoni, C., Salani, S., Fortunato, F., Strazzer, S., Bresolin, N., and Comi, G. P. (2007). Neural stem cells LewisX+ CXCR4+ modify disease progression in an amyotrophic lateral sclerosis model. *Brain* 130, 1289-1305.

[97] Hall, P. E., Lathia, J. D., Miller, N. G., Caldwell, M. A., and Ffrench-Constant, C. (2006). Integrins are markers of human neural stem cells. *Stem Cells* 24, 2078-2084.

[98] Lee, A., Kessler, J. D., Read, T. A., Kaiser, C., Corbeil, D., Huttner, W. B., Johnson, J. E., and Wechsler-Reya, R. J. (2005). Isolation of neural stem cells from the postnatal cerebellum. *Nat. Neurosci.* 8, 723-729.

[99] Tamaki, S., Eckert, K., He, D., Sutton, R., Doshe, M., Jain, G., Tushinski, R., Reitsma, M., Harris, B., Tsukamoto, A., Gage, F., Weissman, I., and Uchida, N. (2002). Engraftment of sorted/expanded human central nervous system stem cells from fetal brain. *J. Neurosci. Res.* 69, 976-986.

[100] Uchida, N., Buck, D. W., He, D., Reitsma, M. J., Masek, M., Phan, T. V., Tsukamoto, A. S., Gage, F. H., and Weissman, I. L. (2000). Direct isolation of human central nervous system stem cells. *Proc. Natl. Acad. Sci. USA* 97, 14720-14725.

[101] Weigmann, A., Corbeil, D., Hellwig, A., and Huttner, W. B. (1997). Prominin, a novel microvilli-specific polytopic membrane protein of the apical surface of epithelial cells, is targeted to plasmalemmal protrusions of non-epithelial cells. *Proc. Natl. Acad. Sci. USA* 94, 12425-12430.

[102] Gritti, A., Parati, E. A., Cova, L., Frolichsthal, P., Galli, R., Wanke, E., Faravelli, L., Morassutti, D. J., Roisen, F., Nickel, D. D., and Vescovi, A. L. (1996). Multipotential stem cells from the adult mouse brain proliferate and self-renew in response to basic fibroblast growth factor. *J. Neurosci.* 16, 1091-1100.

[103] Morshead, C. M., Reynolds, B. A., Craig, C. G., McBurney, M. W., Staines, W. A., Morassutti, D., Weiss, S., and van der, K. D. (1994). Neural stem cells in the adult mammalian forebrain: a relatively quiescent subpopulation of subependymal cells. *Neuron* 13, 1071-1082.

[104] Vescovi, A. L., Reynolds, B. A., Fraser, D. D., and Weiss, S. (1993). bFGF regulates the proliferative fate of unipotent (neuronal) and bipotent (neuronal/astroglial) EGF-generated CNS progenitor cells. *Neuron* 11, 951-966.

[105] Windrem, M. S., Roy, N. S., Wang, J., Nunes, M., Benraiss, A., Goodman, R., McKhann, G. M., and Goldman, S. A. (2002). Progenitor cells derived from the adult human subcortical white matter disperse and differentiate as oligodendrocytes within demyelinated lesions of the rat brain. *J. Neurosci. Res.* 69, 966-975.

In: Glycolipids: New Research
Editor: Daiki Sasaki, pp. 99-116
ISBN: 978-1-60456-216-3

Chapter IV

Mycobacterial Glycolipids and Host Responses

***Nagatoshi Fujiwara**[*1] **and Kazuo Kobayashi**[2]*
[1]Department of Host Defense,
Osaka City University Graduate School of Medicine,
1-4-3 Asahi-machi, Abeno-ku, Osaka 545-8585, Japan
[2]Department of Immunology, National Institute of Infectious Diseases,
1-23-1 Toyama, Shinjuku-ku, Tokyo 162-8640, Japan

Abstract

Mycobacterial diseases, including tuberculosis, leprosy, and disease due to nontuberculous mycobacteria, are the major cause of death from infectious diseases around the world. About one-third of the world population is latently infected with *Mycobacterium tuberculosis*. Over 8 million new cases and nearly 2 million deaths occur each year resulted from tuberculosis. Tuberculosis presents a significant health threat to the world. The pathogenicity of mycobacteria is related to their ability to escape killing by ingested macrophages, latent infection, and induce delayed-type hypersensitivity, such as granulomas. This has been attributed to several components of the mycobacterial cell wall, such as, complement activation factor, heat-shock protein, mycobacterial DNA binding protein, and glycolipids, including mycoloyl glycolipids (containing cord factor/trehalose dimycolate), phenolic glycolipid, sulfolipid, glycopeptidolipid, phosphatidylinositol mannoside, lipoarabinomannan, and lipomannan. Among them, glycolipids specific for mycobacteria play a key role in the pathogenesis, because mycobacteria are rich in lipids that possess pleiotropic activities. The complex interaction between a range of mycobacterial components and the host participate the pathogenesis. In this review we will focus on the pleiotropic activity of glycolipids in both the microbe and host. The better understanding of mycobacterial pathogenicity may open the new avenue for the development of diagnostics, therapeutic and prophylactic interventions.

[*] E-mail: fujiwara@med.osaka-cu.ac.jp

Abbreviation

M. avium-intracellulare complex, MAC;
cord factor/trehalose dimycolate, TDM;
phenolic glycolipid, PGL;
sulfolipid, SL;
glycopeptidolipid, GPL;
phosphatidylinositol mannoside, PIM;
lipoarabinomannan, LAM;
lipomannan, LM;
trehalose 6-monomycolate, TMM;
glucose-6-monomycolate, GMM;
interleukin, IL;
tumor necrosis factor, TNF;
interferon, IFN;
vascular endothelial growth factor, VEGF;
mannose-6-monomycolate, MMM;
fructose-6-monomycolate, FMM;
Toll-like receptor, TLR-2

Introduction

Robert Koch had reported that the pathogenic germ of tuberculosis was acid-fast bacteria in 1882 [1]. After that, tuberculosis has been the most serious problem as the leading cause of death from a single infectious agent in the world. About one-third of the world population is latently infected with *Mycobacterium tuberculosis*, which is one of acid-fast bacteria. Over 8 million new cases and nearly 2 million deaths occur each year resulted from tuberculosis [2]. In the members of the genus *Mycobacterium*, *M. tuberculosis*, *M. avium-intracellulare complex* (MAC), and *M. leprae* are the major pathogen in humans. The host defense against mycobacterial infection is finally intensive two aspects. One is anti-microbial defense, and the other is granulomatous inflammation that is a specific type of chronic reaction characterized by accumulations of modified macrophages (epithelioid cells). A granuloma is a focal area of granulomatous inflammation. It consists of a microscopic aggregation of macrophages that are transformed into epithelium-like cells surrounded by a collar of mononuclear leukocytes, principally lymphocytes and occasionally plasma cells. The precise mechanism for the development of disease is not fully understood.

The first step of infection is intracellular survival in the phagocytic cells, and this process is partially regulated by cell surface glycolipids as virulent factors. The most prominent feature of acid-fast bacteria is to possess a numerous kinds of lipid containing glycolipids and phospholipids in the cell envelope, and their amounts occupy 10 to 40% of the total cell, that is very high compare to other pathogenic bacteria [3]. These lipid components are considered to be influence to not only bacterial physiology and morphology but also host responses. In particularly, mycolic acid, α-branched-β-hydorxy fatty acid is a unique in nature, and ubiquitous in *Actinomycetales* containing *Mycobacterium*, *Nocardia*, *Rhodococcus*, *Gordona*, and *Corynebacterium*. The carbon-chain length of mycolic acid is species-specific, and those of mycobacteria are the longest carbon-chain length (C70-90) [4, 5]. It is becoming evident that these lipid components are related to both host responses and bacterial pathogenesis [6-9]. Until now, many characteristic glycolipids derived from mycobacteria have been reported, for example mycoloyl glycolipids (representative cord factor/trehalose dimycolate, TDM), phenolic glycolipid (PGL), sulfolipid (SL), glycopeptidolipid (GPL), phosphatidylinositol mannoside (PIM), lipoarabinomannan (LAM), and lipomannan (LM) [10, 11]. In this review, we focus the major glycolipids of mycobacteria from the aspect of biochemistry, biology, and immunology.

Heterogeneity of Glycolipids and Phospholipids in Mycobacterial Species

The cell envelope of acid-fast bacteria, including mycobacteria, is wax-like, and rich in lipids. Proposed structure of mycobacterial cell envelope is shown in Figure 1 [10-12]. The distribution of glycolipids, but not phospholipids, is heterogeneous among mycobacterial species. The lipid is composed of the virulent (*M. tuberculosis* H37Rv and *M. tuberculosis* Aoyama B) and avirulent (*M. tuberculosis* H37Ra) strains by using two-dimensional thin-layer chromatography (Figure 2 and Table 1). The composition of glycolipids differs markedly in each strain. Some glycolipids, such as cord factor/TDM and trehalose 6-monomycolate (TMM), exist ubiquitously, but the amount was inconstant. Several glycolipids, including SLs and penta-acyl trehalose, exist only in virulent strains. In general, the composition and content of glycolipids, representative SLs, are heterogeneous among mycobacterial species.

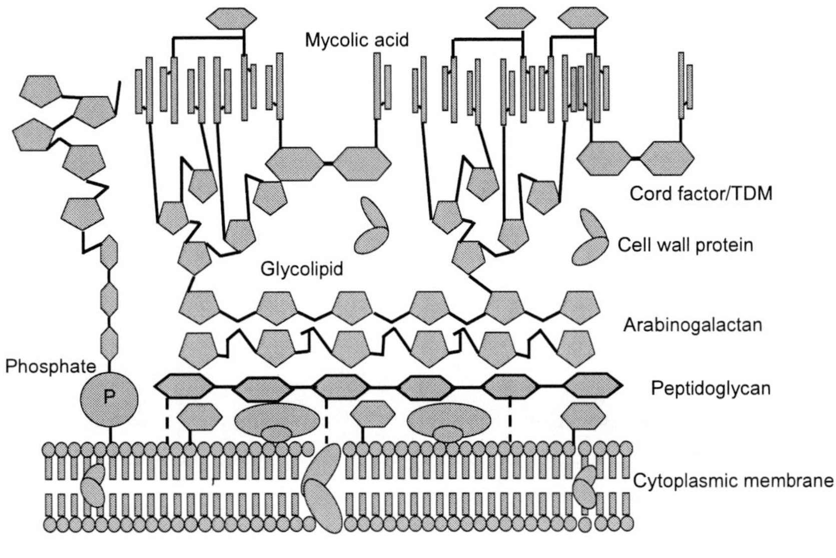

Figure 1. Schematic mycobacterial cell envelop.

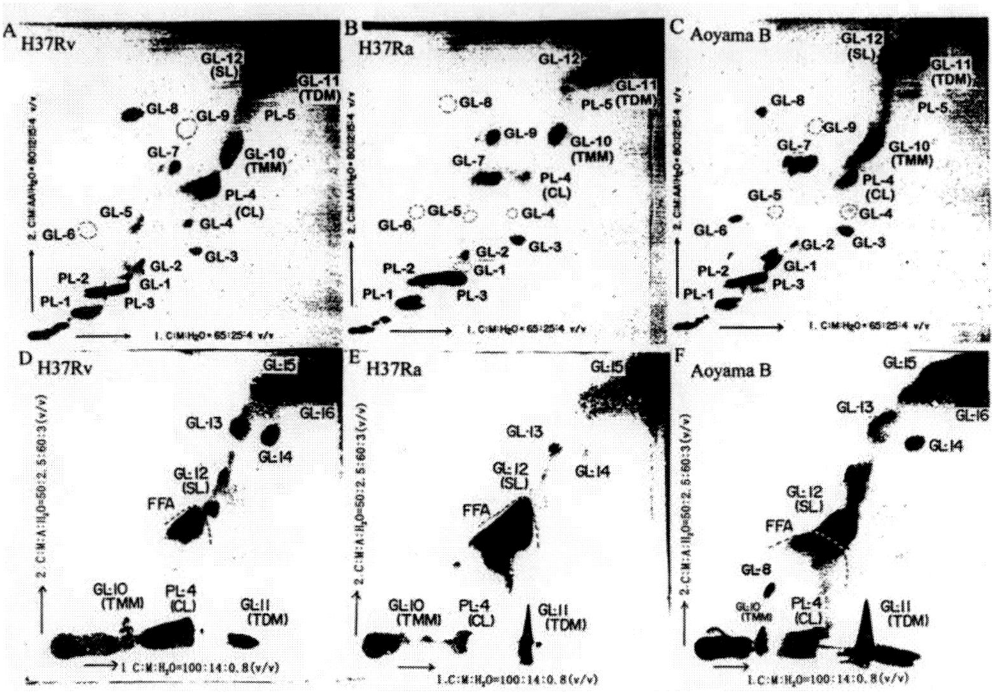

Figure 2. Two dimensional thin-layer chromatography of extractable lipid derived from virulent and avirulent mycobacterial strains. A and D, *M. tuberculosis* H37Rv; B and E, *M. tuberculosis* Aoyama B; C and F, *M. tuberculosis* H37Ra.

The Pathogenicity of Cell Wall Glycolipids

Mycobacteria are the intracellular pathogens. Mycobacterial infection results in dual consequences of the host, development of inflammatory lesions and clearing the pathogen. Genetic regulation and the expression of cell-mediated immunity and delayed-type hypersensitivity play critical roles in the outcome. Cell-mediated immunity participates in host defense, whereas delayed-type hypersensitivity involves in development of granulomatous inflammation. After mycobacterial invasion to the host cell, the bacteria multiply inside macrophages. The mechanism of pathogenesis is focused on escape from host killing system and induction of delayed-type hypersensitivity [13]. Certain cell wall glycolipids, such as cord factor/TDM, SL and LAM [14-16], involve in the host-pathogen interaction, and induce the primary host immune responses. According to current model of the mycobacterial cell envelope (Figure 1), mycolate attached to arabinogalactan forms one highly arranged leaflet of a second membrane, and other lipids participate in the construction of the outer leaflet. This structure composed a very hydrophobic barrier, and is considered to be responsible for resistance to certain drugs.

Table 1. Distribution of phospholipids and glycolipids in strains of *M. tuberculosis*

Strain	Phospholipid						Glycolipid															
	1	2	3	4	5	6	1	2	3	4	5	6	7	8	9	10	11	12	13	14	15	16
H37Rv	+	+	+	+	+	+	+	+	+	+	+	-	+	+	-	+	+	+	+	+	+	+
Aoyama B	+	+	+	+	+	+	+	+	+	-	-	+	+	+	-	+	+	+	+	+	+	+
H37Ra	+	+	+	+	+	+	-	+	+	-	-	-	+	-	+	+	+	-	+	+	+	-

PL-1, 2, phosphatidylinositol mannosides; PL-3, phosphatidyl ethanolamine; PL-4, diphosphatidylglycerol (cardiolipin)

GL-1, di-acyl-trehalose-2'-sulfate (SL-3); GL-8, tri-acyl-trehalose-2'-sulfate (SL-2); GL-10, trehalose-6-monomycolate (TMM); GL-11, trehalose-6,6'-dimycolate (cord factor/TDM); GL-12, penta-acyl-trehalose-2'-sulfate (SL-1); GL-16, penta-acyl-trehalose (PAT). +, present; -, absent

Table 2. Pleiotropic activities of cord factor/TDM derived from *Mycobacterium tuberculosis*

Types of granulomas	Induction of both foreign-body and hypersensitivity granulomas	
Immunogenicity	Expression of cell-mediated immunity	
Angiogenesis	Neovascularization around granulomas	
Apoptosis	Thymocytes, splenocytes	
Chemokines/Cytokines	Chemokines:	
	CXC-chemokine; IL-8	Increased
	CC-chemokines; MCP-1, MIP-1α	Increased
	Proinflammatory; IL-1β, TNF-α	Increased
	Immunoregulatory:	
	Th1-related; IL-12, IFN-γ	Increased
	Th2-related; IL-4	Depressed
	Angiogenic:	
	Vascular endothelial growth factor (VEGF), IL-8	Increased

Table 3. Structurally defined serotype-specific oligosaccharide of GPLs from *M. avium-intracellulare* complex

Serotype	Species	Haptenic oligosaccharde structure
1	*M. avium*	α-L-Rha*p*-(1→2)-L-deoxy-Tal
2	*M. avium*	4-*O*-Ac-2,3-di-*O*-Me-α-L-Fuc*p*-(1→3)-α-L-Rha*p*-(1→2)-L-deoxy-Tal
3	*M. avium*	2,3-di-*O*-Me-α-L-Fuc*p*-(1→4)-ß-D-Glc*p*A-(1→4)-2,3-di-*O*-Me-α-L-Fuc*p*-(1→3)-α-L-Rha*p*-(1→2)-L-deoxy-Tal
4	*M. avium*	4-*O*-Me-α-L-Rha*p*-(1→4)-2-*O*-Me-α-L-Fuc*p*-(1→3)-α-L-Rha*p*-(1→2)-L-deoxy-Tal
7	*M. intracellulare*	4-(2-OH)propionamido-4,6-dideoxy-2-*O*-Me-ß-Hex*p*-(1→3)-α-L-Rha*p*-(1→3)-α-L-Rha*p*-(1→3)-α-L-Rha*p*-(1→2)-L-6-deoxy-Tal
8	*M. avium*	4,6-*O*-(1-carboxyethylidene)-3-*O*-Me-ß-D-Glc*p*-(1→3)-α-L-Rha*p*-(1→2)-L-deoxy-Tal
9	*M. avium*	4-*O*-Ac-2,3-di-*O*-Me-α-L-Fuc*p*-(1→4)-ß-D-Glc*p*A-(1→4)-2,3-di-*O*-Me-α-L-Fuc*p*-(1→3)-α-L-Rha*p*-(1→2)-L-deoxy-Tal
12	*M. intracellulare*	4-(2-OH)propionamido-4,6-dideoxy-3-*O*-Me-ß-D-Glc*p*-(1→3)-4-*O*-Me-α-L-Rha*p*-(1→3)-α-L-Rha*p*-(1→3)-α-L-Rha*p*-(1→2)-L-deoxy-Tal
14	*M. intracellulare*	4-formamido-4,6-dideoxy-2-*O*-Me-3-*C*-Me-α-L-Man*p*-(1→3)-2-*O*-Me-α-D-Rha*p*-(1→3)-α-L-Rha*p*-(1→2)-L-deoxy-Tal
17	*M. intracellulare*	3-(3-OH-2-*O*-Me)butanamido-3,6-dideoxy-ß-D-Glc*p*-(1→3)-4-*O*-Me-α-L-Rha*p*-(1→3)-α-L-Rha*p*-(1→3)-α-L-Rha*p*-(1→2)-L-deoxy-Tal
19	*M. intracellulare*	3,4-di-*O*-Me-ß-D-Glc*p*A-(1→3)-3-*C*-Me-2,4-di-*O*-Me-α-L-Rha*p*-(1→3)-α-L-Rhap-(1→3)-α-L-Rha*p*-(1→2)-L-deoxy-Tal
20	*M. intracellulare*	2-*O*-Me-α-D-Rha*p*-(1→3)-2-*O*-Me-α-L-Fuc*p*-(1→3)-α-L-Rha*p*-(1→2)-L-deoxy-Tal
21	*M. avium*	4,6-*O*-(1-carboxyethylidene)-ß-D-Glc*p*-(1→3)-α-L-Rha*p*-(1→2)-L-deoxy-Tal
25	*M. intracellulare*	2-*O*-Me-α-D-Fuc*p*NAc-(1→4)-ß-D-Glc*p*A-(1→4)-2-*O*-Me-α-L-Fuc*p*-(1→3)-α-L-Rha*p*-(1→2)-L-deoxy-Tal
26	*M. avium*	2,4-di-*O*-Me-α-L-Fuc*p*-(1→4)-ß-D-Glc*p*A-(1→4)-2-*O*-Me-α-L-Fuc*p*-(1→3)-α-L-Rha*p*-(1→2)-L-deoxy-Tal

Peptides and proteins are recognized by host immune system with a wide range of antigenic moieties using major histocompatibility complex (MHC) class I or II. The recognition of lipids is important for host defense against mycobacterial infection as well as other antigenic responses. Several bacterial lipid antigens were recognized by specific T cells. Mycolic acid derived from mycobacteria is lipid antigen that stimulates CD1b-restricted T cells [17, 18]. As other immunogenic mycobacterial glycolipids, glucose-6-monomycolate (GMM), PIM, and LAM are available for loading onto CD1 molecule and recognized by specific T cells [19, 20].

Cord Factor/TDM and Mycoloyl Glycolipids

Cord factor/TDM is prominent mycoloyl glycolipid and first isolated as toxic substance of virulent *M. tuberculosis*. After that, mycoloyl glycolipids have been found widely in acid-fast bacteria and exert potent biological and immunological activities. It has been clarified the unique structure of these mycoloyl glycolipids (Figure 3) and the pleiotropic activities, including the development of granulomatous inflammation, anti-tumor immune response, and adjuvant effects based on the induction of proinflammatory and type 1 helper T cell (Th1)-related cytokines from host cells [21-23].

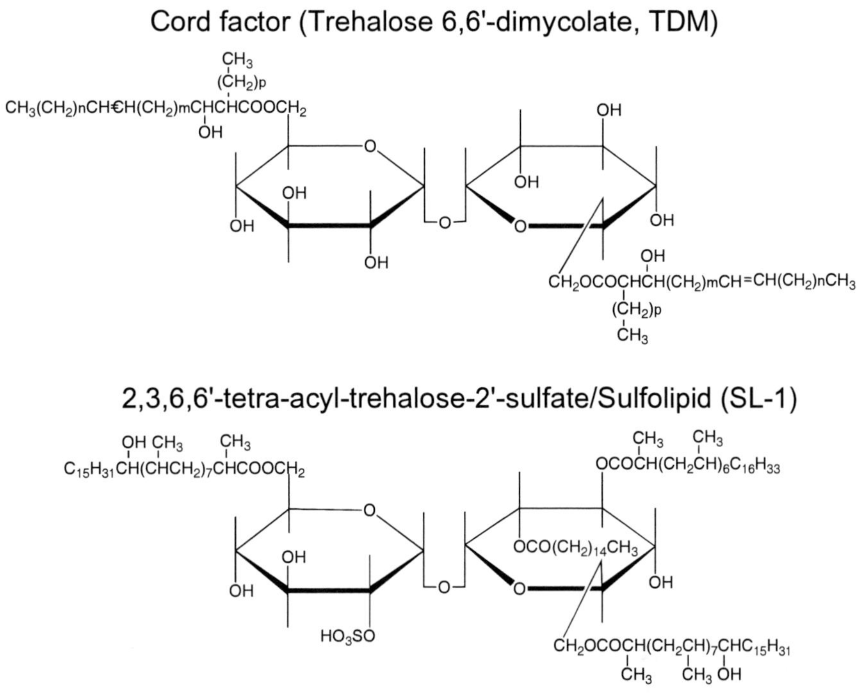

Figure 3. Proposed structures of cord factor/TDM and sulfolipid.

Cord factor/TDM is surface component of tubercle bacilli, and may contact with host in the early stage on infection. Administration of cord factor/TDM can induce delayed-type hypersensitivity-like lesions and foreign-body granulomas in euthymic and athymic mice, regardless of preimmunization with mycobacteria and cord factor/TDM. However, preimmunized mice challenged with cord factor/TDM developed more severe lesions than unimmunized mice. At the active lesion, it was found cc-chemokines (i.e., monocyte chemoattractant protein 1 and macrophage inflammatory protein 1) attracting monocytes, proinflammatory cytokines (i.e., interleukin [IL]-1 and tumor necrosis factor [TNF]-α), and immunoregulatory cytokines (i.e., IL-12 and interferon [IFN]-γ). The inflammatory and cytokine responses were augmented in immunized mice challenged with cord factor/TDM. These results implies that cord factor/TDM can induce both foreign body-type (nonimmune) and hypersensitivity-type (immune) granulomas by acting as a nonspecific irritant and T-cell-dependent antigen, and both nonimmune and immune mechanisms participate in granulomatous inflammation induced by mycobacterial infection [13, 24].

Neovascularization/angiogenesis is required for the progression of chronic inflammation [25]. Although the mechanism of inflammatory neovascularization in tuberculosis remains unknown, cord factor/TDM augments vascular endothelial growth factor (VEGF) production from neutrophils and macrophages, and induces neovascularization in the local granulomatous tissue induced by cord factor/TDM [26, 27]. Apoptosis contributes to the host defense against infection with intracellular bacterial pathogens, and can influence disease progression [28]. Cord factor/TDM also induces atrophy of the thymus and spleen via the apoptotic process, which is closely linked to granuloma formation [29, 30]. Cytokine responses to mycobacterial cord factor/TDM orchestrate the complex event, including granulomatous inflammation, angiogenesis, apoptosis, and probably host defense (Table 2 and Figure 4). Although the sensitivity and specificity are insufficient, mycoloyl glycolipids are employed as an antigen for serodiagnosis of mycobacterial disease, including tuberculosis [31-34].

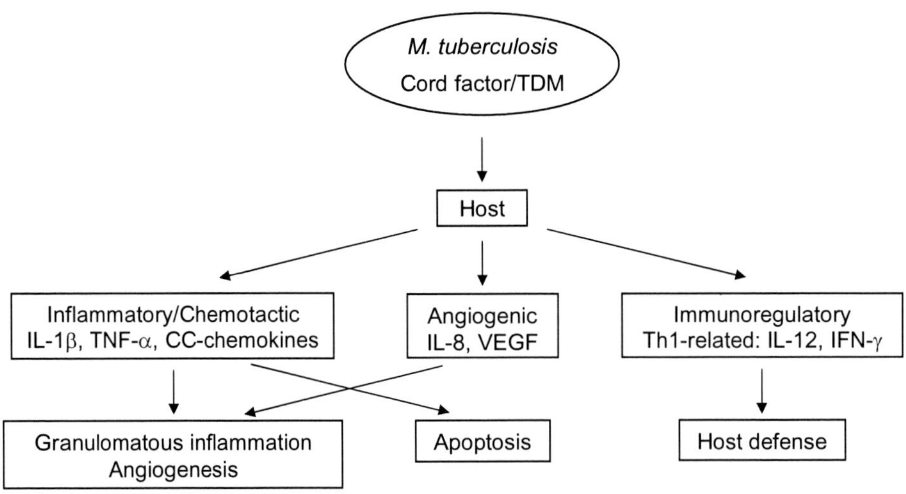

Figure 4. Orchestration of host defense against mycobacterial cord factor/TDM.

Therefore, cord factor/TDM plays a critical role in the pathogenesis of tuberculosis, and is a pleiotropic molecule against the host and participates in the pathogenesis of tuberculosis.

Recently, it is clarified that the modification of mycolic acids influences the pathogenicity in addition to bacterial morphology. The deletion of the proximal cyclopropane ring of α-mycolic acid or of methoxy- and keto-mycolic acids in *M. tuberculosis* results in a significant attenuation in growth when mice were infected [6-8, 35].

In addition to other mycoloyl glycolipids, GMM, mannose-6-monomycolate (MMM), and fructose-6-monomycolate (FMM) are characteristic components of mycobacteria, and those structures are composed of mycolic acid attached to a single glucose, mannose, and fructose residues at C6 position, respectively [36]. These mycololy glycolipids are synthesized when cultivating with each carbon source. GMM shows stronger granuloma-forming and chemotactic activities compared to MMM and FMM [37], although the activity of cord factor/TDM and TMM is more potent. GMM is recognized by CD1b-restricted T cells as antigen [20, 38]. Because this antigenic GMM is synthesized by pathogenic mycobacteria that use host-derived glucose, the immune response against GMM might be distinguished from ubiquitous, but innoculous, environmental mycobacteria.

Sulfolipid (SL)

Mycobacterial SLs are classified into three types by the number of attached short acyl chains. SL-1 is the predominant and SL-2 and 3 are intermediate type. The structure of SL-1 was identified as 2-palmitoyl(stearoyl)-3-phthioceranoyl-6,6'-bis-hydroxyphthioceranoyl-trehalose-2'-sulfate (Figure 3) [39-42]. The presence of SLs is specific for virulent mycobacteria against humans, and is considered to be one of the virulence factors. Several studies have demonstrated the significant relationship between SL-1 and virulence, such as the amount of SL-1 biosynthesis in virulent strains, and the inhibition of phagosome-lysosome fusion in macrophages by SL-1. SL-1 modulates superoxide release and secretion of IL-1β and TNF-α by blocking activation of human macrophages and neutrophils [43-45]. In contrast to these studies, it has been reported that *pks2* disruption and SL-1 deficiency do not significantly affect the replication, persistence, and pathogenicity of *M. tuberculosis* in mice, guinea pigs, or cultured macrophages [46, 47]. The pathogenicity of SL-1 is controversial. SL-1 cannot induce granulomatous inflammation, but rather inhibits granulomatous inflammation and release of TNF-α induced by cord factor/TDM [48]. SL-1 could contribute to virulence at early stage of mycobacterial infection by counteracting the immunopotentiating effect of cord factor/TDM. On the other hand, it is reported that SL-3, 2-palmitoyl(stearoyl)-3-hydroxyphthioceranoyl-trehalose-2'-sulfate, is mainly recognized by CD1b-restricted T cells as a lipid antigen. The other tetra-acylated and tri-acylated SLs (SL-1 and -2) were unable to stimulate the di-acylated SL (SL-3)-specific T cell clones, and implied that immunogenic SL-3 are not generated inside the antigen presented cells [49].

SLs of *M. tuberculosis* have been implicated in the virulence of this organism by inhibiting the phagosome-lysosome fusion in macrophages, thus probably promoting the intracellular survival of *M. tuberculosis*, but the pathogenic role remains controversial. The gene, *pks2*, responsible for SL synthesis was identified and disrupted [50]. The *pks2* mutant defective in an early step in SL biosynthesis, had no obvious growth defect in infected mice. By contrast, a strain lacking MmpL 8, a transporter of SL in *M. tuberculosis*, was highly attenuated for growth in a mouse model of tuberculosis [46]. Although initial replication rates and containment levels of the MmpL 8 mutant were identical, compared with the wild type, a significant attenuation of the mutant strain in time-to-death was observed. Early in infection, differential expression of cytokines and cytokine receptors revealed that the mutant strain less efficiently suppresses key indicators of a Th1-type immune response, suggesting an immunomodulatory role for SLs in the pathogenesis of tuberculosis [51].

Glycopeptidolipid (GPL)

The GPLs are produced by MAC, *M. scrofulaceuem, M. chelonae, M. fortuitum*, and *M. smegmatis*. The polar GPLs produced by MAC species are of particular interest because of their serotype-specific antigenicity [12]. Structurally, GPLs contain a tetrapeptide-amino alcohol core, D-phenylalanine-D-*allo*-threonine-D-alanine-L-alaninol, with an amido-linked 3-hydroxy or 3-methoxy C26-C34 fatty acid at the *N*-terminal of D-Phenylalanine [52]. The D-*allo*-threonine and terminal L-alaninol are further linked with 6-deoxy-talose and 3,4-di-*O*-methyl rhamnose, respectively. This GPL core is common to all serotypes and exhibits an antigenicity [10]. The serotype-specific GPLs are further glycosylated with a variable haptenic oligosaccharide at 6-deoxxy-talose. At present, 31 distinct serotype-specific polar GPLs have been identified biochemically, and the complete structures of 15 GPLs are defined (Table 3) [12, 53]. The standard technique to classify MAC strains has been employed serologic typing based on the oligosaccharide residue of the GPL. More recently, the biosynthetic pathway of various haptenic oligosaccharide is explored and the genes encoding the pathways have been identified and characterized [54-56]. GPL present on the cell wall is considered to be influenced the colony morphology. Changes in the MAC colony phenotype are spontaneously occurred from smooth to rough type, and this is due to a mutation lacking GPL [57-59].

We have demonstrated the availability of serodiagnosis of MAC pulmonary diseases using the GPL and GPL core antigens and have also shown that the levels of GPL and GPL core antibodies reflect disease activity [60-62]. GPL is one of the immunologically active molecules characteristic of MAC, and serotype-specific GPLs participate in the pathogenesis and immunomodulation in the host [63, 64]. It has been reported that the GPL core seems to play a role in suppression of mitogen-induced blastogenic response in spleen cells [65]. In addition, the immunomodulating activity of GPL on macrophage functions is serotype-dependent [66]. The serotype 4 GPL promotes phagocytosis and inhibits phagosome-lysosome (P-L) fusion, whereas the GPLs of serotypes 9 and 16 exhibit no effect on phagocytosis and P-L fusion. The serotype 8 GPL shows concomitant stimulation of both phagocytosis and P-L fusion. Because GPL core, but not oligosaccharides, is common in all serotypes, the oligosaccharide of GPL may be involved in the mechanism of inhibition of P-L fusion, which is mediated through mannose receptors of macrophages [67]. The serotype 4 GPL inhibited lymphoproliferative response to mitogens [64]. Thus, host responses to GPLs vary with the MAC serotype. It is reported that the uptake by and growth in macrophages of a MAC mutant inactivated the gene of GPL synthesis pathway by transposon insertion were decreased [68]. The pathogenicity of GPL may comprise both a common peptide core and an oligosaccharide elongated from 6-deoxy-talose. GPL is a pleiotropic molecule and participates in the pathogenesis of MAC disease. Elucidation of the structure-activity relationship of GPL is required to better understand the pathogenesis.

Phenolic Glycolipids (PGL)

The glycosylated phenolphthiocerol dimycocerosates, so-called phenolic glycolipid (PGL) is produced by a limited group of mycobacterial species, and most of them are pathogenic for humans [69]. PGL is distributed in *M. leprae* as a unique antigen, and inhibits the lymphoproliferative responses and suppress monocyte oxidative responses [70-72]. It has also shown that disruption of PGL synthesis results in loss of virulent phenotype without significantly affecting bacterial load during disease in experimental models using mice and rabbits, and loss of PGL was found to correlate with an increase in the release of the pro-inflammatory cytokines, TNF-α, IL-6, and IL-12 in vitro [73, 74].

A PGL purified from *M. leprae* has been used as antigen in an enzyme-linked immunosorbent assay. Antibodies directed against the lipid were detected in sera of leprosy patients but not in sera from uninfected controls or patients infected with other mycobacteria, including *M. tuberculosis*. The antibody response distinguished between the *M. leprae* lipid and the structurally related PGL from *M. kansasii*. Similar to serodiagnosis of MAC disease using GPL, this assay based on PGL antigen has considerable potential as a specific serodiagnostic test for infection with *M. leprae* [75-78].

Phosphatidylinositol Mannoside (PIM), Lipomannan (LM), and Lipoarabinomannan (LAM)

PIMs and their multiglycosylated counterparts, LM and LAM, are complex lipoglycans that are found ubiquitously in the envelopes of all mycobacterial species. Their structures originate from a phosphatidyl-*myo*-inositol (MPI) anchor, which is mannosylated to generate LM and further arabinosylated to give LAM [79]. LM is the biosynthetic precursor of LAM. The non-reducing termini of the arabinosyl side chains can be substituted by capping motifs, and LAM classified into three families. LAM from slow growing mycobacteria bearing mannose caps, *i.e.* mono- or (α1→2)-di- or tri-mannoside units, are designated as ManLAM. In contrast, LAMs from fast growing mycobacteria capped by phospho-*myo*-inositol units and not capped at all, are termed PILAM and AraLAM, respectively [80]. LAM and LM exhibit a broad spectrum of immunomodulatory activities, including the ability to modulate the production of macrophage-derived Th1 pro-inflammatory cytokines, most commonly TNF-α and IL-12. ManLAM from *M. tuberculosis* is involved in the inhibition of phagosome maturation, apoptosis and IFN-γ signaling in macrophages and IL-12 secretion of dendritic cells (DCs). ManLAM contributes an immunosuppressive effect to the persistence of slow-growing mycobacteria in the humans. In contrast, PILAMs are able to induce the release of a variety of proinflammatory cytokines. Recently it is reported that LMs from both pathogenic and nonpathogenic mycobacterial species, independent of their origin, are potent stimulators of TNF-α, IL-8, and IL-12, and activate macrophages via Toll-like receptor (TLR)-2 [81-83]. The ManLAM/LM balance might be important to host immune responses against mycobacteria. In addition, LAM-based vaccine for tuberculosis is currently under investigation using experimental guinea pig models [84].

PIM is composed of a phsophatidylinositol and mannose moieties. PIMs have 4-6 mannoses and the third and fourth of them are acylated in some cases. PIM_2 containing two mannoses is highly immunogenic and a main component of mycobacterial cell walls. Several biological functions have been recently attributed to PIM. PIM_2 are shown to recruit natural killer T cells, which play a role in the local granulomatous response [85, 86]. PIM activates cells via TLR-2 [87]. PIM_6 as well as ManLAM from *M. leprae* and *M. tuberculosis* are presented by antigen-presenting cells in the context of CD1b [19]. The phosphatidylinositol moiety plays a central role in the process of PIM and ManLAM binding to CD1b proteins.

Concluding Remarks

Mycobacterial glycolipids are pleiotropic molecules and play a key role in both microbes and hosts acting as structural components of cell walls and immunologically active substances. A better understanding of mycobacterial glycolipids helps us to develop novel diagnostics, therapeutics, and prophylactics for mycobacterial diseases.

Grant Support

This work was supported by the grants from Ministry of Health, Labour and Welfare (Emerging and Re-emerging Infectious Diseases), Ministry of Education, Culture, Sports, Science and Technology of Japan, and Japan Health Sciences Foundation.

References

[1] R. Koch. (1884). *Aetiologie der Tuberkulose*. Mitt. k. Gesundheitsamt, 2, 1-88.

[2] WHO. (2007). WHO report 2007 Global Tuberculosis Control Surveillance, Planning, Financing. http://www.who.int/tb/publications/global_report/2007/pdf/full.pdf,

[3] M. Daffe and P. Draper. (1998). The envelope layers of mycobacteria with reference to their pathogenicity. *Adv. Microb. Physiol.*, 39, 131-203.

[4] K. Kaneda, S. Imaizumi and I. Yano. (1995). Distribution of C22-, C24- and C26-alpha-unit-containing mycolic acid homologues in mycobacteria. *Microbiol. Immunol.*, 39, 563-570.

[5] K. Kaneda, S. Imaizumi, S. Mizuno, T. Baba, M. Tsukamura and I. Yano. (1988). Structure and molecular species composition of three homologous series of alpha-mycolic acids from *Mycobacterium* spp. *J. Gen. Microbiol.*, 134, 2213-2229.

[6] M. S. Glickman, J. S. Cox and W. R. Jacobs, Jr. (2000). A novel mycolic acid cyclopropane synthetase is required for cording, persistence, and virulence of *Mycobacterium tuberculosis*. *Mol. Cell.*, 5, 717-727.

[7] V. Rao, F. Gao, B. Chen, W. R. Jacobs, Jr. and M. S. Glickman. (2006). Trans-cyclopropanation of mycolic acids on trehalose dimycolate suppresses *Mycobacterium tuberculosis* -induced inflammation and virulence. *J. Clin. Invest.*, 116, 1660-1667.

[8] E. Dubnau, J. Chan, C. Raynaud, V. P. Mohan, M. A. Laneelle, K. Yu, A. Quemard, I. Smith and M. Daffe. (2000). Oxygenated mycolic acids are necessary for virulence of *Mycobacterium tuberculosis* in mice. *Mol. Microbiol.*, 36, 630-637.

[9] Y. Yuan, Y. Zhu, D. D. Crane and C. E. Barry, 3rd. (1998). The effect of oxygenated mycolic acid composition on cell wall function and macrophage growth in *Mycobacterium tuberculosis*. *Mol. Microbiol.*, 29, 1449-1458.

[10] G. O. Aspinall, D. Chatterjee and P. J. Brennan. (1995). The variable surface glycolipids of mycobacteria: structures, synthesis of epitopes, and biological properties. *Adv. Carbohydr. Chem. Biochem.*, 51, 169-242.

[11] P. J. Brennan and H. Nikaido. (1995). The envelope of mycobacteria. *Annu. Rev. Biochem.*, 64, 29-63.

[12] D. Chatterjee and K. H. Khoo. (2001). The surface glycopeptidolipids of mycobacteria: structures and biological properties. *Cell Mol. Life Sci.*, 58, 2018-2042.

[13] K. Kobayashi and T. Yoshida. (1996). The immunopathogenesis of granulomatous inflammation induced by *Mycobacterium tuberculosis*. *Methods*, 9, 204-214.

[14] M. S. Glickman and W. R. Jacobs, Jr. (2001). Microbial pathogenesis of *Mycobacterium tuberculosis*: dawn of a discipline. *Cell*, 104, 477-485.

[15] D. G. Russell, H. C. Mwandumba and E. E. Rhoades. (2002). *Mycobacterium* and the coat of many lipids. *J. Cell. Biol.*, 158, 421-426.

[16] P. J. Brennan. (2003). Structure, function, and biogenesis of the cell wall of *Mycobacterium tuberculosis*. *Tuberculosis* (Edinb), 83, 91-97.

[17] E. M. Beckman, S. A. Porcelli, C. T. Morita, S. M. Behar, S. T. Furlong and M. B. Brenner. (1994). Recognition of a lipid antigen by CD1-restricted alpha beta+ T cells. *Nature*, 372, 691-694.

[18] D. B. Moody, S. A. Porcelli, S. A. Porcelli and R. L. Modlin. (2003). Intracellular pathways of CD1 antigen presentation. *Nat. Rev. Immunol.*, 3, 11-22.

[19] P. A. Sieling, D. Chatterjee, S. A. Porcelli, T. I. Prigozy, R. J. Mazzaccaro, T. Soriano, B. R. Bloom, M. B. Brenner, M. Kronenberg, P. J. Brennan and et al. (1995). CD1-restricted T cell recognition of microbial lipoglycan antigens. *Science*, 269, 227-230.

[20] D. B. Moody, M. R. Guy, E. Grant, T. Y. Cheng, M. B. Brenner, G. S. Besra and S. A. Porcelli. (2000). CD1b-mediated T cell recognition of a glycolipid antigen generated from mycobacterial lipid and host carbohydrate during infection. *J. Exp. Med.*, 192, 965-976.

[21] Y. Natsuhara, S. Oka, K. Kaneda, Y. Kato and I. Yano. (1990). Parallel antitumor, granuloma-forming and tumor-necrosis-factor-priming activities of mycoloyl glycolipids from Nocardia rubra that differ in carbohydrate moiety: structure-activity relationships. Cancer Immunol. *Immunother.*, 31, 99-106.

[22] I. Yano, I. Tomiyasu, K. Kaneda, Y. Kato, Y. Sumi, S. Kurano, N. Sugimoto and H. Sawai. (1987). Isolation of mycolic acid-containing glycolipids in *Nocardia rubra* and their granuloma forming activity in mice. *J. Pharmacobiodyn.*, 10, 113-123.

[23] K. Kaneda, Y. Sumi, F. Kurano, Y. Kato and I. Yano. (1986). Granuloma formation and hemopoiesis induced by C36-48-mycolic acid-containing glycolipids from *Nocardia rubra*. *Infect. Immun.*, 54, 869-875.

[24] H. Yamagami, T. Matsumoto, N. Fujiwara, T. Arakawa, K. Kaneda, I. Yano and K. Kobayashi. (2001). Trehalose 6,6'-dimycolate (cord factor) of *Mycobacterium tuberculosis* induces foreign-body- and hypersensitivity-type granulomas in mice. *Infect. Immun.*, 69, 810-815.

[25] J. R. Jackson, M. P. Seed, C. H. Kircher, D. A. Willoughby and J. D. Winkler. (1997). The codependence of angiogenesis and chronic inflammation. *FASEB J.*, 11, 457-465.

[26] I. Sakaguchi, N. Ikeda, M. Nakayama, Y. Kato, I. Yano and K. Kaneda. (2000). Trehalose 6,6'-dimycolate (Cord factor) enhances neovascularization through vascular endothelial growth factor production by neutrophils and macrophages. *Infect. Immun.*, 68, 2043-2052.

[27] N. Saita, N. Fujiwara, I. Yano, K. Soejima and K. Kobayashi. (2000). Trehalose 6,6'-dimycolate (cord factor) of *Mycobacterium tuberculosis* induces corneal angiogenesis in rats. *Infect. Immun.*, 68, 5991-5997.

[28] L. Y. Gao and Y. A. Kwaik. (2000). The modulation of host cell apoptosis by intracellular bacterial pathogens. *Trends Microbiol.,* 8, 306-313.

[29] Y. Ozeki, K. Kaneda, N. Fujiwara, M. Morimoto, S. Oka and I. Yano. (1997). In vivo induction of apoptosis in the thymus by administration of mycobacterial cord factor (trehalose 6,6'-dimycolate). *Infect. Immun.*, 65, 1793-1799.

[30] N. Hamasaki, K. Isowa, K. Kamada, Y. Terano, T. Matsumoto, T. Arakawa, K. Kobayashi and I. Yano. (2000). In vivo administration of mycobacterial cord factor (Trehalose 6, 6'-dimycolate) can induce lung and liver granulomas and thymic atrophy in rabbits. *Infect. Immun.*, 68, 3704-3709.

[31] R. Maekura, M. Nakagawa, Y. Nakamura, T. Hiraga, Y. Yamamura, M. Ito, E. Ueda, S. Yano, H. He, S. Oka and et al. (1993). Clinical evaluation of rapid serodiagnosis of pulmonary tuberculosis by ELISA with cord factor (trehalose-6,6'-dimycolate) as antigen purified from *Mycobacterium tuberculosis*. *Am. Rev. Respir. Dis.*, 148, 997-1001.

[32] E. Julian, L. Matas, A. Perez, J. Alcaide, M. A. Laneelle and M. Luquin. (2002). Serodiagnosis of tuberculosis: comparison of immunoglobulin A (IgA) response to sulfolipid I with IgG and IgM responses to 2,3-diacyltrehalose, 2,3,6-triacyltrehalose, and cord factor antigens. *J. Clin. Microbiol.*, 40, 3782-3788.

[33] Y. Fujita, H. Ogata and I. Yano. (2005). Clinical evaluation of serodiagnosis of active tuberculosis by multiple-antigen ELISA using lipids from *Mycobacterium bovis* BCG Tokyo 172. *Clin. Chem. Lab. Med.*, 43, 1253-1262.

[34] K. R. Steingart, M. Henry, S. Laal, P. C. Hopewell, A. Ramsay, D. Menzies, J. Cunningham, K. Weldingh and M. Pai. (2007). Commercial serological antibody detection tests for the diagnosis of pulmonary tuberculosis: a systematic review. *PLoS Med.,* 4, e202.

[35] V. Rao, N. Fujiwara, S. A. Porcelli and M. S. Glickman. (2005). *Mycobacterium tuberculosis* controls host innate immune activation through cyclopropane modification of a glycolipid effector molecule. *J. Exp. Med.*, 201, 535-543.

[36] S. Ueda, N. Fujiwara, T. Naka, I. Sakaguchi, Y. Ozeki, I. Yano, T. Kasama and K. Kobayashi. (2001). Structure-activity relationship of mycoloyl glycolipids derived from *Rhodococcus* sp. 4306. *Microb. Pathog.*, 30, 91-99.

[37] I. Matsunaga, S. Oka, N. Fujiwara and I. Yano. (1996). Relationship between induction of macrophage chemotactic factors and formation of granulomas caused by mycoloyl glycolipids from *Rhodococcus ruber* (*Nocardia rubra*). *J. Biochem.* (Tokyo), 120, 663-670.

[38] Y. Enomoto, M. Sugita, I. Matsunaga, T. Naka, A. Sato, T. Kawashima, K. Shimizu, H. Takahashi, Y. Norose and I. Yano. (2005). Temperature-dependent biosynthesis of glucose monomycolate and its recognition by CD1-restricted T cells. *Biochem. Biophys. Res. Commun.*, 337, 452-456.

[39] J. D. Mougous, C. J. Petzold, R. H. Senaratne, D. H. Lee, D. L. Akey, F. L. Lin, S. E. Munchel, M. R. Pratt, L. W. Riley, J. A. Leary, J. M. Berger and C. R. Bertozzi. (2004). Identification, function and structure of the mycobacterial sulfotransferase that initiates sulfolipid-1 biosynthesis. *Nat. Struct. Mol. Biol.*, 11, 721-729.

[40] M. B. Goren, O. Brokl, B. C. Das and E. Lederer. (1971). Sulfolipid I of *Mycobacterium tuberculosis*, strain H37RV. Nature of the acyl substituents. *Biochemistry,* 10, 72-81.

[41] M. B. Goren. (1970). Sulfolipid I of *Mycobacterium tuberculosis*, strain H37Rv. I. Purification and properties. *Biochim. Biophys. Acta*, 210, 116-126.

[42] M. B. Goren. (1970). Sulfolipid I of *Mycobacterium tuberculosis*, strain H37Rv. II. Structural studies. *Biochim. Biophys. Acta*, 210, 127-138.

[43] L. Zhang, D. English and B. R. Andersen. (1991). Activation of human neutrophils by *Mycobacterium tuberculosis*-derived sulfolipid-1. *J. Immunol.*, 146, 2730-2736.

[44] M. Kato and M. B. Goren. (1974). Synergistic action of cord factor and mycobacterial sulfatides on mitochondria. *Infect. Immun.*, 10, 733-741.

[45] M. J. Pabst, J. M. Gross, J. P. Brozna and M. B. Goren. (1988). Inhibition of macrophage priming by sulfatide from *Mycobacterium tuberculosis*. *J. Immunol.*, 140, 634-640.

[46] S. E. Converse, J. D. Mougous, M. D. Leavell, J. A. Leary, C. R. Bertozzi and J. S. Cox. (2003). MmpL8 is required for sulfolipid-1 biosynthesis and *Mycobacterium tuberculosis* virulence. *Proc. Natl. Acad. Sci. USA.,* 100, 6121-6126.

[47] C. Rousseau, O. C. Turner, E. Rush, Y. Bordat, T. D. Sirakova, P. E. Kolattukudy, S. Ritter, I. M. Orme, B. Gicquel and M. Jackson. (2003). Sulfolipid deficiency does not affect the virulence of *Mycobacterium tuberculosis* H37Rv in mice and guinea pigs. *Infect. Immun.*, 71, 4684-4690.

[48] Y. Okamoto, Y. Fujita, T. Naka, M. Hirai, I. Tomiyasu and I. Yano. (2006). Mycobacterial sulfolipid shows a virulence by inhibiting cord factor induced granuloma formation and TNF-alpha release. *Microb. Pathog.*, 40, 245-253.

[49] M. Gilleron, S. Stenger, Z. Mazorra, F. Wittke, S. Mariotti, G. Bohmer, J. Prandi, L. Mori, G. Puzo and G. De Libero. (2004). Diacylated sulfoglycolipids are novel

mycobacterial antigens stimulating CD1-restricted T cells during infection with *Mycobacterium tuberculosis. J. Exp. Med.*, 199, 649-659.

[50] T. D. Sirakova, A. K. Thirumala, V. S. Dubey, H. Sprecher and P. E. Kolattukudy. (2001). The *Mycobacterium tuberculosis pks2* gene encodes the synthase for the hepta- and octamethyl-branched fatty acids required for sulfolipid synthesis. *J. Biol. Chem.*, 276, 16833-16839.

[51] P. Domenech, M. B. Reed, C. S. Dowd, C. Manca, G. Kaplan and C. E. Barry, 3rd. (2004). The role of MmpL8 in sulfatide biogenesis and virulence of *Mycobacterium tuberculosis. J. Biol. Chem.,* 279, 21257-21265.

[52] P. J. Brennan and M. B. Goren. (1979). Structural studies on the type-specific antigens and lipids of the *Mycobacterium avium. Mycobacterium intracellulare. Mycobacterium scrofulaceum serocomplex. Mycobacterium intracellulare* serotype 9. *J. Biol. Chem.*, 254, 4205-4211.

[53] N. Fujiwara, N. Nakata, S. Maeda, T. Naka, M. Doe, I. Yano and K. Kobayashi. (2007). Structural Characterization of a Specific Glycopeptidolipid Containing a Novel *N*-Acyl-Deoxy Sugar from *Mycobacterium intracellulare* Serotype 7 and Genetic Analysis of Its Glycosylation Pathway. *J. Bacteriol.*, 189, 1099-1108.

[54] T. M. Eckstein, J. T. Belisle and J. M. Inamine. (2003). Proposed pathway for the biosynthesis of serovar-specific glycopeptidolipids in *Mycobacterium avium* serovar 2. *Microbiology*, 149, 2797-2807.

[55] T. Heidelberg and O. R. Martin. (2004). Synthesis of the glycopeptidolipid of *Mycobacterium avium* Serovar 4: first example of a fully synthetic C-mycoside GPL. *J. Org. Chem.*, 69, 2290-2301.

[56] J. N. Maslow, V. R. Irani, S. H. Lee, T. M. Eckstein, J. M. Inamine and J. T. Belisle. (2003). Biosynthetic specificity of the rhamnosyltransferase gene of *Mycobacterium avium* serovar 2 as determined by allelic exchange mutagenesis. *Microbiology*, 149, 3193-3202.

[57] S. T. Howard, E. Rhoades, J. Recht, X. Pang, A. Alsup, R. Kolter, C. R. Lyons and T. F. Byrd. (2006). Spontaneous reversion of *Mycobacterium abscessus* from a smooth to a rough morphotype is associated with reduced expression of glycopeptidolipid and reacquisition of an invasive phenotype. *Microbiology,* 152, 1581-1590.

[58] J. T. Belisle, K. Klaczkiewicz, P. J. Brennan, W. R. Jacobs, Jr. and J. M. Inamine. (1993). Rough morphological variants of *Mycobacterium avium*. Characterization of genomic deletions resulting in the loss of glycopeptidolipid expression. *J. Biol. Chem.*, 268, 10517-10523.

[59] T. M. Eckstein, J. M. Inamine, M. L. Lambert and J. T. Belisle. (2000). A genetic mechanism for deletion of the ser2 gene cluster and formation of rough morphological variants of *Mycobacterium avium. J. Bacteriol.*, 182, 6177-6182.

[60] K. Enomoto, S. Oka, N. Fujiwara, T. Okamoto, Y. Okuda, R. Maekura, T. Kuroki and I. Yano. (1998). Rapid serodiagnosis of *Mycobacterium avium-intracellulare* complex infection by ELISA with cord factor (trehalose 6, 6'-dimycolate), and serotyping using the glycopeptidolipid antigen. *Microbiol. Immunol.*, 42, 689-696.

[61] S. Kitada, R. Maekura, N. Toyoshima, N. Fujiwara, I. Yano, T. Ogura, M. Ito and K. Kobayashi. (2002). Serodiagnosis of pulmonary disease due to *Mycobacterium avium*

complex with an enzyme immunoassay that uses a mixture of glycopeptidolipid antigens. *Clin. Infect. Dis.*, 35, 1328-1335.

[62] S. Kitada, R. Maekura, N. Toyoshima, T. Naka, N. Fujiwara, M. Kobayashi, I. Yano, M. Ito and K. Kobayashi. (2005). Use of glycopeptidolipid core antigen for serodiagnosis of *Mycobacterium avium* complex pulmonary disease in immunocompetent patients. *Clin. Diagn. Lab. Immunol.,* 12, 44-51.

[63] W. W. Barrow, T. L. Davis, E. L. Wright, V. Labrousse, M. Bachelet and N. Rastogi. (1995). Immunomodulatory spectrum of lipids associated with *Mycobacterium avium* serovar 8. *Infect. Immun.*, 63, 126-133.

[64] H. Kano, T. Doi, Y. Fujita, H. Takimoto, I. Yano and Y. Kumazawa. (2005). Serotype-specific modulation of human monocyte functions by glycopeptidolipid (GPL) isolated from *Mycobacterium avium* complex. *Biol. Pharm. Bull.*, 28, 335-339.

[65] S. K. Tassell, M. Pourshafie, E. L. Wright, M. G. Richmond and W. W. Barrow. (1992). Modified lymphocyte response to mitogens induced by the lipopeptide fragment derived from *Mycobacterium avium* serovar-specific glycopeptidolipids. *Infect. Immun.,* 60, 706-711.

[66] Y. Takegaki. (2000). [Effect of serotype specific glycopeptidolipid (GPL) isolated from *Mycobacterium avium* complex (MAC) on phagocytosis and phagosome-lysosome fusion of human peripheral blood monocytes]. *Kekkaku*, 75, 9-18.

[67] K. Shimada, H. Takimoto, I. Yano and Y. Kumazawa. (2006). Involvement of mannose receptor in glycopeptidolipid-mediated inhibition of phagosome-lysosome fusion. *Microbiol. Immunol.*, 50, 243-251.

[68] L. Danelishvili, M. Wu, B. Stang, M. Harriff, S. Cirillo, J. Cirillo, R. Bildfell, B. Arbogast and L. E. Bermudez. (2007). Identification of *Mycobacterium avium* pathogenicity island important for macrophage and amoeba infection. *Proc. Natl. Acad. Sci. USA.*, 104, 11038-11043.

[69] M. Daffe and M. A. Laneelle. (1988). Distribution of phthiocerol diester, phenolic mycosides and related compounds in mycobacteria. *J. Gen. Microbiol.,* 134, 2049-2055.

[70] J. J. Fournie, E. Adams, R. J. Mullins and A. Basten. (1989). Inhibition of human lymphoproliferative responses by mycobacterial phenolic glycolipids. *Infect. Immun.*, 57, 3653-3659.

[71] V. Mehra, P. J. Brennan, E. Rada, J. Convit and B. R. Bloom. (1984). Lymphocyte suppression in leprosy induced by unique M. leprae glycolipid. *Nature*, 308, 194-196.

[72] M. Vachula, T. J. Holzer and B. R. Andersen. (1989). Suppression of monocyte oxidative response by phenolic glycolipid I of *Mycobacterium leprae*. *J. Immunol.,* 142, 1696-1701.

[73] M. B. Reed, P. Domenech, C. Manca, H. Su, A. K. Barczak, B. N. Kreiswirth, G. Kaplan and C. E. Barry, 3rd. (2004). A glycolipid of hypervirulent tuberculosis strains that inhibits the innate immune response. *Nature*, 431, 84-87.

[74] L. Tsenova, E. Ellison, R. Harbacheuski, A. L. Moreira, N. Kurepina, M. B. Reed, B. Mathema, C. E. Barry, 3rd and G. Kaplan. (2005). Virulence of selected *Mycobacterium tuberculosis* clinical isolates in the rabbit model of meningitis is dependent on phenolic glycolipid produced by the bacilli. *J. Infect. Dis.,* 192, 98-106.

[75] D. B. Young and T. M. Buchanan. (1983). A serological test for leprosy with a glycolipid specific for *Mycobacterium leprae*. *Science*, 221, 1057-1059.

[76] D. B. Young, S. Dissanayake, R. A. Miller, S. R. Khanolkar and T. M. Buchanan. (1984). Humans respond predominantly with IgM immunoglobulin to the species-specific glycolipid of *Mycobacterium leprae*. *J. Infect. Dis*., 149, 870-873.

[77] S. M. van Beers, S. Izumi, B. Madjid, Y. Maeda, R. Day and P. R. Klatser. (1994). An epidemiological study of leprosy infection by serology and polymerase chain reaction. *Int. J. Lepr. Other Mycobact. Dis.,* 62, 1-9.

[78] S. Izumi, T. Fujiwara, M. Ikeda, Y. Nishimura, K. Sugiyama and K. Kawatsu. (1990). Novel gelatin particle agglutination test for serodiagnosis of leprosy in the field. *J. Clin. Microbiol.,* 28, 525-529.

[79] S. W. Hunter and P. J. Brennan. (1990). Evidence for the presence of a phosphatidylinositol anchor on the lipoarabinomannan and lipomannan of *Mycobacterium tuberculosis*. *J. Biol. Chem*., 265, 9272-9279.

[80] J. Nigou, M. Gilleron and G. Puzo. (2003). Lipoarabinomannans: from structure to biosynthesis. *Biochimie.,* 85, 153-166.

[81] D. N. Dao, L. Kremer, Y. Guerardel, A. Molano, W. R. Jacobs, Jr., S. A. Porcelli and V. Briken. (2004). *Mycobacterium tuberculosis* lipomannan induces apoptosis and interleukin-12 production in macrophages. *Infect. Immun*., 72, 2067-2074.

[82] V. J. Quesniaux, D. M. Nicolle, D. Torres, L. Kremer, Y. Guerardel, J. Nigou, G. Puzo, F. Erard and B. Ryffel. (2004). Toll-like receptor 2 (TLR2)-dependent-positive and TLR2-independent-negative regulation of proinflammatory cytokines by mycobacterial lipomannans. *J. Immunol.,* 172, 4425-4434.

[83] C. Vignal, Y. Guerardel, L. Kremer, M. Masson, D. Legrand, J. Mazurier and E. Elass. (2003). Lipomannans, but not lipoarabinomannans, purified from *Mycobacterium chelonae* and *Mycobacterium kansasii* induce TNF-alpha and IL-8 secretion by a CD14-toll-like receptor 2-dependent mechanism. *J. Immunol*., 171, 2014-2023.

[84] Y. Watanabe, E. Watari, I. Matsunaga, K. Hiromatsu, C. C. Dascher, T. Kawashima, Y. Norose, K. Shimizu, H. Takahashi, I. Yano and M. Sugita. (2006). BCG vaccine elicits both T-cell mediated and humoral immune responses directed against mycobacterial lipid components. *Vaccine,* 24, 5700-5707.

[85] M. Gilleron, C. Ronet, M. Mempel, B. Monsarrat, G. Gachelin and G. Puzo. (2001). Acylation state of the phosphatidylinositol mannosides from *Mycobacterium bovis* bacillus Calmette Guerin and ability to induce granuloma and recruit natural killer T cells. *J. Biol. Chem.,* 276, 34896-34904.

[86] I. Apostolou, Y. Takahama, C. Belmant, T. Kawano, M. Huerre, G. Marchal, J. Cui, M. Taniguchi, H. Nakauchi, J. J. Fournie, P. Kourilsky and G. Gachelin. (1999). Murine natural killer T(NKT) cells [correction of natural killer cells] contribute to the granulomatous reaction caused by mycobacterial cell walls. *Proc. Natl. Acad. Sci. USA*., 96, 5141-5146.

[87] B. W. Jones, T. K. Means, K. A. Heldwein, M. A. Keen, P. J. Hill, J. T. Belisle and M. J. Fenton. (2001). Different Toll-like receptor agonists induce distinct macrophage responses. *J. Leukoc. Biol*., 69, 1036-1044.

In: Glycolipids: New Research
Editor: Daiki Sasaki, pp. 117-131
ISBN: 978-1-60456-216-3

Chapter V

Antiglycolipid Antibodies in Autoimmune Neuropathies; New Aspects of Research

Susumu Kusunoki*
Department of Neurology,
Kinki University School of Medicine, Japan

Abstract

In autoimmune neuropathies such as Guillain-Barré syndrome (GBS) and IgM paraproteinemic neuropathy, serum antiglycolipid antibodies are frequently elevated in titer. Those antibodies are useful diagnostic markers and some of them may be directly involved in the pathogenetic mechanisms. In GBS, the carbohydrate structures of the microorganisms of the antecedent infections that mimic those of human glycolipids may induce production of the antiglycolipid antibodies. "Molecular mimicry" mechanism has been shown in the production of antiganglioside antibodies after *Campylobacter jejuni* infection and antigalactocerebroside antibodies after *Mycoplasma pneumoniae* infection. The antibodies may specifically bind to the regions where the respective antigens are located (for example, the antibodies against the ganglioside GQ1b may cause ophthalmoplegia by binding to the paranodal regions of the oculomotor, trochlear, and abducens nerves, where GQ1b is densely localized). We have proved this hypothesis by inducing sensory ataxic neuropathy in rabbits sensitized with the ganglioside GD1b, which is localized in the primary sensory neurons conveying deep sensation. We have recently found that the antibodies that specifically recognize a new conformational epitope formed by two gangliosides are sometimes present in the acute-phase sera of GBS patients. In particular, the antibodies against GD1a-GD1b complex are associated with severe GBS requiring artificial ventilation. Gangliosides are preferentially packaged with cholesterol to form lipid rafts, in which the carbohydrate portions of two different gangliosides may form a new conformational epitope. Within the rafts, gangliosides are

* Corresponding Author: Susumu Kusunoki; 377-2 Ohno-Higashi, Osaka-Sayama, Osaka 589-8511, Japan; Tel: +81-72-366-0221; Fax: +81-72-368-4846; e-mail: kusunoki-tky@umin.ac.jp

considered to interact with important receptors or signal transducers. The antibodies against ganglioside complex may therefore directly induce nerve conduction failure and severe disability in GBS. More study is needed to elucidate the roles of antiglycolipid antibodies in the pathogenetic mechanisms of autoimmune neuropathies.

Introduction

In immune-mediated neuropathies such as Guillain-Barré syndrome (GBS) and IgM paraproteinemic neuropathy, antiglycolipid antibodies, which specifically recognize the carbohydrate portion of the glycolipids, are frequently present in the patients' sera. There are diversities in the carbohydrate sequences of the glycolipids (figure 1). Each glycolipid has a unique distribution within the peripheral nervous sysytem. Those antibodies can be used as useful markers for diagnosis. In addition, considering that the glycolipids are localized in the plasma membrane with their carbohydrate portions extended to the extracellular spaces, they may be involved in the pathogenetic process.

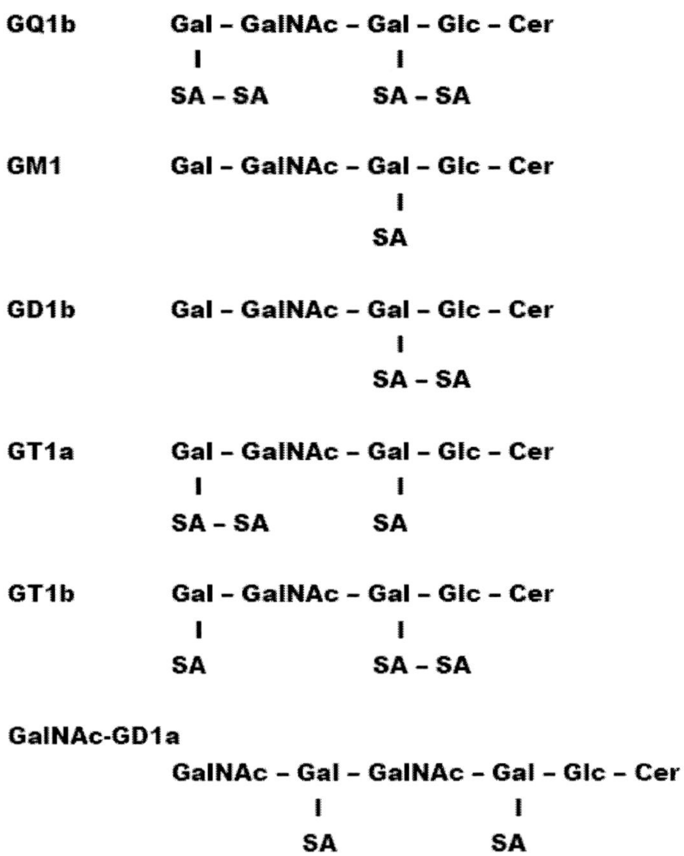

Figure 1. Carbohydrate sequences of gangliosides; Gal: galactose, GalNAc: N-acetylgalactosamine, Glc: glucose; SA: sialic acid, Cer: ceramide, SA-SA: disialosyl residue.

1. Guillain-Barré Syndrome

GBS is an acute, self-limited motor-dominant polyneuropathy. An infection, involving respiratory or gastrointestinal tract commonly precedes the neurological onset. GBS is considered as an autoimmne disease, of which the target is either the myelin or the axon of the peripheral nervous systems. The former is referred to as acute inflammatory demyelinating polyneuropathy (AIDP) and the latter acute axonal motor neuropathy (AMAN)[1]. Diagnosis of GBS is made by the characteristic clinical course, signs, and symptoms. Electrophysiological and cerebrospinal fluid examination, elevation of protein content but not of cell count (alubuminocytological dissociation), have been known to be useful for diagnosis. Recent serological investigations using a panel of the glycolipid antigens revealed that approximately 60% of sera from patients with acute-phase GBS had antibodies, predominantly IgG class, against at least one glycolipid[2,3]. Gangliosides, which are sialic acid-containing glycolipids and are distributed in the nervous system, are frequently the target antigens for serum antibodies in GBS. The antibody titers are highest in the acute phase and decrease with clinical improvement. In contrast, elevation of protein in cerebrospinal fluid usually becomes obvious on 7-10 days of neurological onset. It suggests that elevation of the antiglycolipid antibody titer may not be a result of the damage to the peripheral nervous system but may be associated with the pathogenetic mechanisms. Microorganisms of the antecedent infections of GBS sometimes are shown to have carbohydrate structures similar to those of such glycolipids as gangliosides. Molecular mimicry has therefore been suggested as the mechanism of producing antiglycolipid antibodies. Antiglycolipid antibodies should be useful for diagnosis and also may be important pathogenetic factors of GBS. Several antibodies associated with GBS and related disorders will be described as below.

a) Anti-GQ1b Antibody

Miller Fisher syndrome (MFS) is a variant of GBS, characterized by the triad of ophthalmoplegia, ataxia, and areflexia. It shares such features as frequent presence of antecedent infection, acute and self-limited clinical course, and albuminocytological dissociation in cerebrospinal fluid with GBS. Specific and frequent (more than 90%) elevation of anti-GQ1b IgG antibody titers in MFS was reported in 1992[4]. The titers of this antibody are also elevated in the acute-phase sera of GBS with ophthalmoplegia[5], "atypical MFS" (ophthalmoplegia without ataxia[5] or ataxia without ophthalmoplegia[6] of acute post-infectious self-limited clinical course), and "Bickerstaff type brainstem encephalitis[7], which has clinical characteristics of MFS and central nervous system involvement. On the other hand, this antibody is not detected in sera from patients with other diseases. The elevation of serum anti-GQ1b IgG antibody titers is therefore closely associated with the occurrence of acute ophthalmoplegia and/or ataxia caused by such pathogenetic mechanisms as those underlying GBS. The antibody assay has been utilized for differentiating this syndrome from other immunological or neurological diseases, such as multiple sclerosis and myasthenia gravis, even when ophthalmoplegia is present.

GQ1b has two disialosyl residues attached to the gangliotetraose structure (Gal-GalNAc-Gal-Glc-). Although both GT1a and GT1b have carbohydrate sequences similar to that of GQ1b, most of the MFS sera have IgG antibody activity against GT1a as well as GQ1b but not against GT1b[5]. An absorption study showed that the same antibody in MFS sera binds to both GQ1b and GT1a[5]. The serum antibodies in MFS sera may therefore recognize the disialosyl residue attached to the outer galactose of the gangliotetraose structure.

Biochemical analysis of ganglioside fractions extracted from cranial nerves and the ventral and dorsal roots showed that GQ1b is present in any of those nerves although the oculomotor, trochlear and abducens nerves had a relatively higher content of GQ1b[8]. However, an immunohistochemical study of the human nervous system using a mouse monoclonal antibody (7F5) provided us with a reason why anti-GQ1b antibody is closely associated with ophthalmoplegia and ataxia. As the monoclonal antibody 7F5 recognizes not only GQ1b but also GT1a, it has the same binding specificity as the IgG antibodies in MFS sera. The results showed that the paranodal myelin of the extramedullary portion of the cranial nerves innervating the extraocular muscles (oculomotor, trochlear and abducens nerves) was specifically immunostained[5]. No significant staining was observed in other cranial or peripheral nerves. The paranodal region is essential to nerve conduction and is the site of the earliest pathological change in GBS. Specific binding of the anti-GQ1b IgG to the paranodal myelin of the cranial nerves innervating extraocular muscles may underlie the pathogenesis of ophthalmoplegia in MFS and GBS (figure 2). The monoclonal antibody 7F5 also immunostained a subset of neurons in the dorsal root ganglia and the cerebellar granule cells of humans [6]. Binding of the anti-GQ1b IgG antibody to those sites may be associated with the development of ataxia in MFS. Considering the looseness of the blood-nerve barrier in DRG, DRG neurons may be more accessible for the antibody in serum. A neurophysiological study suggested the dysfunction of the proprioceptive afferent system in MFS [9]. If the afferent fibers from GQ1b-positive DRG neurons connect with the spinocerebellar tract, the anti-GQ1b IgG antibodies may be involved in the development of ataxia by binding to those GQ1b-positive neurons.

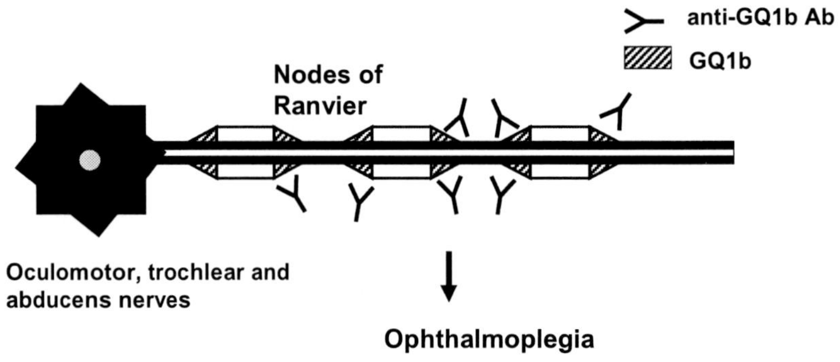

Figure 2. Specific binding of the anti-GQ1b IgG to the paranodal myelin of the cranial nerves innervating extraocular muscles may cause ophthalmoplegia in MFS and GBS.

It was reported that anti-GQ1b antibodies bind to neuromuscular junctions (NMJs), cause massive quantal release of acetylcholine from nerve terminals and eventually block neuromuscular transmission in mouse diaphragm[10]. This resembles the effect of alpha-latrotoxin. It was also reported that morphological destruction of the motor nerve terminal in mouse hemidiaphragm preparation was induced by anti-GQ1b antibody-positive MFS serum and a mouse monoclonal anti-GQ1b antibody[11]. Those experimental results definitively showed that anti-GQ1b IgG antibodies can potentially cause motor weakness by their action to NMJs. However, the close association between that antibody and ophthalmoplegia may be difficult to explain solely based on those results, because the effect observed in the mouse diaphragma preparation should occur not only in the cranial nerves innervating the ocular muscles but also in the other peripheral nerves. MFS patients sometimes have other cranial nerve palsies. Anti-GQ1b IgG antibodies are also present in patients who have prominent limb weakness as well as ophthalmoplegia (GBS with ophthalmoplegia) as described above. Among GBS patients, it was reported that anti-GQ1b IgG-positive patients more frequently needed assistance with aritificial ventilation[12]. The effect of the anti-GQ1b IgG antibodies on NMJs could be involved in the pathogenesis of general motor weakness that sometimes occur in patients with anti-GQ1b IgG antibodies. Abnormal sudomotor axon reflex in patients with anti-GQ1b IgG antibodies has recently been reported, which suggests that the anti-GQ1b antibody also has similar effects on the neuroglandular junction[13].

Most of the antecedent infections of MFS or GBS with anti-GQ1b IgG antibodies are respiratory tract infections, although approximately 10 % are gastrointestinal infections, in which *Campylobacter jejuni* is the frequent infectious agent. *C jejuni* from stools from anti-GQ1b IgG positive patients has GQ1b-like carbohydrate structure in the lipopolysaccharide (LPS), indicating the antibody produced in response to the GQ1b-like epitope of the bacterial LPS is reactive with GQ1b ("molecular mimicry" mechanism)[14,15].

Antibodies against ganglioside-complexes including GQ1b will be discussed below.

b) Anti-GM1 Antibody

Anti-GM1 antibodies are one of the most frequent antiganglioside antibodies in GBS, detected in approximately 30 % of acute-phase GBS patients' sera[16]. Anti-GM1 IgG antibodies were reported to be associated with severe AMAN[17,18]. In subsequent reports, however, they were shown to be present not only in severe cases but in AMAN with mild clinical course [19]. In addition, other researchers reported that they are detected not only in sera of AMAN but also in those of AIDP[20]. Further investigation by collaboration of several research groups using definite neurophysiological criteria is needed for clarification of the clinical significance of anti-GM1 IgG antibodies in GBS.

It has been reported that GM1 is localized in the nodes of Ranvier and paranodal regions in the peripheral nervous system in studies using cholera toxin and peanut agglutinin lectin [21]. Those probes however bind to several gangliosides as well as GM1. Although GM1 is present in human PNS, specific localization of GM1 in human PNS has not been determined by the use of such probes monospecific to GM1 as the monoclonal antibody GMB16 [22].

Sensitization of rabbits with GM1 was shown to induce motor axonal neuropathy. IgG was deposited on the axons of the ventral roots[23]. This experimental model suggests the important role of anti-GM1 antibody in the pathogenesis of motor axonal neuropathy.

Both gastrointestinal and respiratory tract infections can precede GBS with anti-GM1 IgG antibodies. Among the former, *C. jejuni* is the most frequent infectious agent. *C. jejuni* isolated from the stool of the anti-GM1 IgG-positive GBS patients has the carbohydrate structure similar to that of GM1, indicating anti-GM1 antibody is produced by the "molecular mimicry" mechanism [24].

c) Anti-GD1b Antibody

Two gangliosides, GM1 and GD1b, share the terminal galactosyl beta1-3 N-acetylgalactosaminyl (Gal-GalNAc-) residue. Antibodies recognizing the Gal-GalNAc-residue, therefore, bind to both GM1 and GD1b. Antibodies reactive with both GM1 and GD1b are present in approximately 20% of acute-phase GBS sera, anti-GM1 without anti-GD1b reactivity in approximately 10%, and anti-GD1b without anti-GM1 reactivity in approximately 10%[16]. Among the last group, there are monospecific anti-GD1b antibodies, which are reactive with only GD1b but not with other gangliosides or glycolipids. To elucidate the clinical significance of anti-GD1b reactivity, we should investigate the clinical features of GBS patients with monospecific anti-GD1b IgG antibodies.

Among 445 patients with GBS, 70 had anti-GD1b IgG antibodies. Nine of those patients had monospecific anti-GD1b IgG antibodies. Eight of the nine patients had respiratory tract infections before neurological onset. All the nine patients had sensory disturbances. Electrophysiological studies showed that none of the nine patients had AMAN form[25]. In the human PNS, GD1b is localized in paranodal myelin of both motor and sensory nerves as well as primary sensory neurons[22]. Anti-GD1b IgG antibodies may bind to these regions and cause sensory disturbances and demyelination. There are several case reports of GBS with anti-GD1b antibodies in which ataxia is prominent. Rabbits sensitized with GD1b were affected with ataxic neuropathy with high titer anti-GD1b antibodies, as detailed in later[26]. Only 3 of the above 9 patients however had ataxia[25]. Fine specificity of each monospecific anti-GD1b IgG antibodies may be varied and further investigation is necessary to clarify this point.

d) Anti-Galnac-GD1a Antibodies

GalNAc-GD1a is a relatively minor ganglioside which migrates just below GD1a in thin-layer chromatogram (TLC). Anti-GalNAc-GD1a IgG antibodies are reported to be present in about 10% of acute-phase GBS sera[16]. Monospecific anti-GalNAc-GD1a IgG antibodies are closely associated with the pure motor variant of GBS characterized by distal-dominant weakness and infrequent cranial nerve involvement[27]. Subsequent immunohistochemical investigation using a monospecific rabbit antibody showed that the antibody immunostained the periaxonal-axolemma-related region or an inner part of compact myelin of motor nerve fibers[28]. Some sensory nerve fibers however are also immunostained the rabbit anti-GalNAc-GD1a antibody. The reason for the discrepancy between the immunolocalization of GalNAc-GD1a in sensory nerves and the absence of sensory disturbance in patients with

GBS with IgG anti-GalNAc-GD1a antibodies is not known. Anti-GalNAc-GD1a IgM antibodies which are cross-reactive with GM2 are present in sera from GBS with sensory disturbances[29]. The anti-GalNAc-GD1a antibodies with such specificity might bind to GalNAc-GD1a in sensory fibers to cause sensory disturbances, whereas the monospecific anti-GalNAc-GD1a IgG might not. Future investigation is necessary to clarify this point.

e) Antigalactocerebroside Antibodies

Galactocerebroside (Gal-C) is a major glycolipid antigen in the myelin of both the central and peripheral nervous systems. Sensitization of rabbits with Gal-C causes demyelinating neuropathy, and anti-Gal-C antibody was shown to be a demyelinating factor[30]. Although elevation of anti-Gal-C antibody titers is not so common in GBS patients, the antibodies are quite frequently present in GBS subsequent to *Mycoplasma pneumoniae* infection[31]. Anti-Gal-C antibody activities in these sera were inhibited specifically by the *M. pneumoniae* reagent. A rabbit anti-Gal-C antibody recognized several glycolipids in the lipid fraction extracted from *M. pneumoniae*. Although their carbohydrate structure remains to be shown, those glycolipids in M. pneumoniae that are recognized by anti-Gal-C antibody may stimulate the immune system to produce anti-Gal-C antibody ("molecular mimicry mechanism"), which may participate in the pathogenesis of GBS as a demyelinating factor[32].

f) Antibodies to other Gangliosides

Antibodies to such gangliosides as GD1a, GT1b, LM1, GM2, and GM1b have also been reported in the acute-phase sera of GBS[2,3]. Anti-GD1a IgG[33] and anti-GM1b IgG antibodies are reported to be related with AMAN. Patients with anti-LM1 IgG antibodies frequently have anti-GQ1b IgG antibodies as well[34]. Anti-GM2 (IgM and/or IgG) antibodies are known to be associated with preceding cytomegalovirus infection[35](table 1).

Table 1. Clinical features and antiglycolipid antibodies in GBS

Target antigen	class	localization of antigen	Clinical features
GQ1b	IgG	paranodal myelin of cranial nerves III,IV,VI some neurons in DRG	Miller Fisher syndrome GBS with ophthalmoplegia Bickerstaff brainstem encephalitis
GM1	IgG	not determined	AMAN Pure motor GBS
GD1b (monospecific)	IgG	paranodal myelin large neurons in DRG	AIDP ataxic GBS
GalNAc-GD1a	IgG	periaxonal membrane	AMAN Pure motor GBS
Galactocerebroside	IgG, IgM	myelin	AIDP
GM2	IgG, IgM	not determined	cranial nerve involvement

g) Antibodies to Ganglioside Complexes

i) Anti-GD1a/GD1b Complex Antibodies

Gagliosides have characteristics of forming clusters in the plasma membrane[36]. In the clusters, the carbohydrate structure of a ganglioside may intereact with each other to form a novel epitope. Clustered glycoepitopes of ganglioside complexes in the membrane may be targeted by serum antibodies in GBS patients. We recently revealed that some GBS patients had serum antibodies that specifically recognize the novel epitope formed by two ganglioside molecules and named such antibodies as "anti-ganglioside complex antibodies"[37].

We first investigated a serum from a 31-year-old man who developed acute flaccid tetraparesis after several days of flu-like symptoms. GBS was diagnosed based on Asbury and Cornblath criteria. His acute phase serum was checked for anti-ganglioside antibodies by an enzyme-linked immunosorbent assay (ELISA) and thin-layer chromatogram (TLC) immunostaining. We first found an unidentified immuno-reactive band in the position just below GD1a on TLC of a crude ganglioside fraction from bovine brain. ELISA results were negative for each of the test gangliosides (GalNAc-GD1a, GM1, GM2, GM3, GD1a, GD1b, GD3, GT1b, and GQ1b). But the serum IgG bound strongly to the well coated with the mixture of GD1a and GD1b gangliosides (GD1a/GD1b). In TLC-immunostaining using a developing solvent of chloroform/methanol/0.2%$CaCl_2$-$2H_2O$ (50:45:10), positive staining was present in the lane in which both GD1a and GD1b were developed, but not in the lanes in which GD1a or GD1b was developed respectively (figure 3). In another developing solvent (C/M/0.2%$CaCl_2$-$2H_2O$, 30/65/10) that separated the positions of GD1a and GD1b more than the previous one, the immunostaining disappeared. The Image Analyzer assay showed specific immunostaining in the overlapping portion of the GD1a and GD1b antigens (figure).

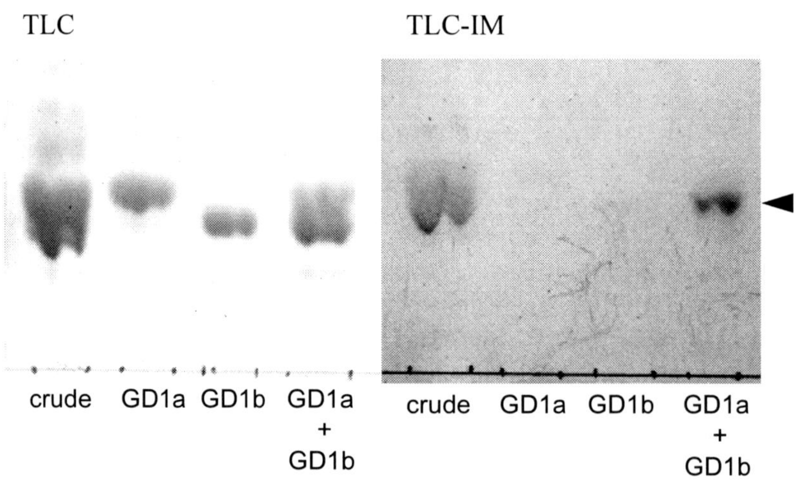

Figure 3. TLC-immunostaining with anti-GD1a/GD1b-positive sera; Positive staining was present in the lane in which both GD1a and GD1b were developed, but not in the lanes in which GD1a or GD1b was developed respectively.

To investigate the frequency of anti-GD1a/GD1b antibody in GBS patients, serum anti-ganglioside antibodies were investigated by ELISAs using 100 consecutive acute-phase GBS sera. Anti-GD1a- or anti-GD1b-positive sera, in which the corrected anti-GD1a/GD1b antibody OD was 0.2 higher than the corrected anti-GD1a or anti-GD1b antibody OD, were considered anti-GD1a/GD1b antibody-positive. The cut-off value (0.2) for the anti-GD1a/GD1b antibody was decided arbitrarily. The results showed that eight out of 100 consecutive patients (8%) with GBS had IgG anti-GD1a/GD1b antibodies. None of the disease or normal control group patients did. Three of the eight GBS patients with IgG anti-GD1a/GD1b antibodies had neither the IgG anti-GD1a nor anti-GD1b antibody. Of 92 GBS patients without IgG anti-GD1a/GD1b antibody, four had IgG anti-GD1a antibody, and eleven had IgG anti-GD1b antibody. TLC immunostaining of the anti-GD1a/GD1b antibody-positive sera gave the same results as for the above patient. Four anti-GD1a/GD1b antibody-positive patients required artificial ventilation (grade 5), and 3 other patients were bed- or chair-bound (grade 4) at peak. Electrophysiological studies showed "primary axonal" in two patients, equivocal" in one patient, which were classified as described elsewhere.[5] Five other patients had no available electrophysiological data.

Anti-GD1a/GD1b antibody-positive sera often showed little or no reactivity to GD1a or GD1b in the ELISAs. And even in the anti-GD1a- or anti-GD1b- positive cases, OD values of the anti-GD1a/GD1b antibodies were much higher than those of the anti-GD1a or anti-GD1b antibodies. It is therefore indicated that mixing GD1a with GD1b produced a ganglioside complex and a new conformational glycoepitope that is different from those of GD1a or GD1b alone.

The anti-GD1a/GD1b antibody-positive sera also had antibody activities to two or more of ganglioside complexes such as GD1a/GM1, GD1b/GT1b, and GM1/GT1b. Future investigation is needed on the presence of antibodies to the various ganglioside complexes using sera from larger number of GBS patients.

Glycosphingolipids are located in the cell membranes with carbohydrate portions on the outer surfaces, and are preferentially packaged with cholesterol, forming lipid rafts. These lipid rafts, protein-linked microdomains in the plasma membrane, are called detergent-insoluble glycolipid-enriched complexes or glycosphingolipid-enriched membranes. Within the plasma membrane microdomains, gangliosides are considered to interact with important transmembrane receptors or signal transducers involved in cell adhesion and signaling[36,38]. Antibodies to a ganglioside complex therefore may have effects on the function of the axon or Schwann cell through their binding to clustered epitopes of glycosphingolipids in the plasma membrane microdomains. Thus, they may directly induce nerve conduction failure and severe disability in patients with GBS. Anti-GD1a/GD1b antibody-positive patients with GBS tend to have severe disabilities. Clinical studies of larger numbers of patients with GBS are needed to clarify the significance of anti-GD1a/GD1b antibody.

ii) Antibodies to Ganglioside Complexes Including GQ1b

As described above, the presence in serum of the IgG anti-GQ1b antibody is an excellent diagnostic marker for MFS[4,5]. Because MFS is considered to be a variant of GBS, MFS patients may also have anti-GSC antibodies. Presence of anti-ganglioside complexes antibodies in MFS therefore was investigated using seven ganglioside antigens; GM1, GM2, GD1a, GD1b, GT1a, GT1b and GQ1b [39].

Acute phase serum samples were collected from 12 MFS patients, 10 (83%) of whom had IgG anti-GQ1b antibodies. ELISA results showed that seven (58%) of the 12 patients had serum antibodies to ganglioside complexes such as GQ1b/GM1, GQ1b/GD1b, GQ1b/GD1a, GT1a/GM1, GT1a/GD1b, GT1a/GD1a and GQ1b/GT1b, but not to GSCs without GQ1b or GT1a. Especially, antibodies to GQ1b/GM1, GT1a/GM1, and GT1a/GD1b were frequent. One patient had no anti-GQ1b or anti-GT1a antibodies, but had antibodies to GQ1b/GM1 and GT1a/GM1. In contrast to GBS, no patients had antibodies to the complexes consisting of two of the four major gangliosides, GM1, GD1a, GD1b and GT1b. TLC-immunostaining also showed specific immunoreactivity against the overlapping portion of two gangliosides, including GT1a/GM1, GQ1b/GM1 and GT1a/GD1b.

The ELISA showed that the fine specificity of serum antibodies in MFS patients was heterogeneous. Some serum antibodies had stronger activity with GQ1b/GM1 or GT1a/GM1 than with either GQ1b or GT1a alone, whereas others had little or no activity against GQ1b/GM1 or GT1a/GM1, despite showing intense activity with GQ1b or GT1a. Based on the results of anti-ganglioside-complex antibody assay, the 12 MFS patients could be subdivided into the three groups; antibody negative for ganglioside complexes (5 patients), anti-GQ1b/GM1-positive (5 patients), and anti-GQ1b/GD1a-positive (2 patients).

Antibodies to GSCs containing GQ1b or GT1a, as well as anti-GQ1b and anti-GT1a antibodies, may be critical for development of MFS. Serum antibodies specific to GQ1b/GM1 and GT1a/GM1 may recognize a combination of [Galβ1-3GalNAc] and [NeuAcα2-8NeuAcα2-3Galβ1-3GalNAc] in the terminal residues of gangliotetraose structures. In contrast, serum antibodies specific to GQ1b/GD1a and GQ1b/GT1b may recognize a combination of [NeuAcα2-3Galβ1-3GalNAc] and [NeuAcα2-8NeuAcα2-3Galβ1-3GalNAc] in the terminal residues. On the other hands, some patients had monospecific activities to GQ1b or GT1a. Thus, there are at least three different specificities in MFS-associated antibodies. Because sensory signs were infrequent in MFS patients with antibodies to GQ1b/GM1 and GT1a/GM1 but were frequent in patients with other antibodies, such differences in antibody specificity may have some influence on clinical features in MFS. However, the clinical relevance of such anti-GSC antibodies needs to be investigated in a larger number of MFS patients before firm conclusions can be drawn.

In view of the characteristic formation of clusters of gangliosides in the plasma membrane, anti-GSC antibodies may cause nerve dysfunction more efficiently than monospecific anti-GQ1b antibody. Future investigations of the localization and possible roles of ganglioside complexes in the plasma membrane are required.

2. IgM Paraproteinemic Neuropathy

IgM paraproteinemia is often associateed with chronic peripheral neuropathies. IgM paraproteins (M-proteins) associated with neuropathies frequently recognize varied carbohydrate epitopes. Each binding specificity of the IgM M-protein is linked with a unique clinical feature [2].

a) IgM M-Protein Specific to Myelin-Associated Glycoprotein, P0-Glycoprotein, and Phosphated Glucuronyl Paragloboside

Approximately half of the IgM M-proteins of such IgM paraproteinemic neuropathies bind to the carbohydrate epitope, sulfated glucuronyl residue, shared by myelin-associated glycoprotein (MAG) and such glycolipids as sulfated glucuronyl paragloboside (SGPG)[40, 41]. MAG and SGPG are localized in myelin. IgM deposits are found in the myelin in sural nerve biopsy specimens. Passive transfer of this IgM M-protein caused demyelination of the peripheral nerve[42]. IgM M-protein of this specificity may play an important role in the pathogenesis of the demyelinating neuropathy.

In addition to MAG and SGPG, P0-glycoprotein and peripheral myelin protein-22 (PMP-22) also have sulfated glucuronyl residue. Some of the IgM M-proteins binding to SGPG strongly immunostained peripheral nerve myelin, whereas the others strongly immunostained the cytoplasm of the Schwann cells surrounding the myelin sheath with only weak staining of the myelin. Western blotting showed that the former IgM M-proteins also reacted strongly with P0 and PMP-22, whereas the latter did not. P0 and PMP-22 are known to be localized in compact myelin. The strong reactivity with P0 and PMP-22 may be associated with the strong immunostaining of the peripheral nerve myelin by the former IgM M-protein[43].

b) IgM M-Protein that Binds to Gangliosides with a Disialosyl Residue

In some patients with IgM paraproteinemic neuropathy, IgM M-protein binds to gangliosides with a disialosyl resudie, such as GD1b, GD3, GT1b and GQ1b. IgM M-protein of this type is specifically associated with sensory ataxic neuropathy[44]. GD1b is localized in the primary sensory neurons with large diameters[22]. The IgM M-protein of this type may therefore bind to GD1b in the primary sensory neurons to cause sensory ataxic neuropathy.

c) Experimental Sensory Ataxic Neuropathy Induced by Sensitization with GD1b

As described above, GD1b ganglioside is a target molecule for serum antibodies in sensory ataxic neuropathy. GD1b is localized in neurons with large diameters in rabbit DRG as well as human DRG. We therefore assumed that sensitization with GD1b may induce sensory ataxic neuropathy in rabbits. As a result, experimental sensory ataxic neuropathy indeed developed in about 50% of the rabbits immunized with GD1b[26]. The affected rabbits lay on the floor with splayed limbs, frequently in a peculiar posture. The limbs moved awkwardly. Muscle power and pain sensation were not affected. Anti-GD1b antibody titers were markedly raised in rabbits immunized with GD1b. Pathological investigation of the affected rabbits showed axonal degeneration in the dorsal root, the dorsal column of the spinal cord, and the sciatic nerve, whereas the ventral root was intact. No demyelination was found. Some nerve cell bodies in DRG had degenerated or disappeared. No lymphocytic infiltration was observed in the affected regions. It is therefore indicated that anti-GD1b antibody may cause specific degeneration of primary sensory neurons mediating deep sensation. This was supported by the report that degeneration of rabbit sensory neurons could be induced by passive transfer of anti-GD1b antiserum[45]. Monospecific anti-GD1b IgG antibody was shown to be required to induce this sensory ataxic neuropathy[46].

DRG neurons with large diameters depend on neurotrophin-3-mediated trkC signaling. TrkC expression was reported to be reduced in DRG of the affected rabbits in the acute phase[47]. Binding of the anti-GD1b antibody may cause downregulation of trkC expression. It may be involved in one of the pathogenetic mechanisms of this animal model.

This is the first established animal model of autoimmune neuropathy mediated by the antiganglioside antibody. This model confirms the assumption that antiganglioside antibodies may determine the clinical features of autoimmune neuropathies by binding to the respective gangliosides that have a unique distribution in the peripheral nervous system.

Conclusion

Antiglycolipid antibodies are useful diagnostic markers in autoimmune neuropathies. Recent investigations showed that they are also involved directly in the pathogenetic mechanisms. Different species of gangliosides may interact with each other to form a novel carbohydrate epitope, which may also be target antigens for serum antibodies in autoimmune neuropathies. Future investigation on antiglycolipid antibodies to clarify the pathogenetic mechanisms of autoimmune neuropathies may also be useful to understand the physiological functions of glycolipids in the nervous system.

References

[1] Asbury AK, McKhann GM. Changing views of Guillain-Barré syndrome. *Ann. Neurol.* 1997; 41: 287-288.

[2] Kusunoki S. Antiglycolipid antibodies in Guillain-Barré syndrome. *Am. J. Med. Sci.* 2000; 319: 234-239.

[3] Willison HJ, Yuki N. Peripheral neuropathies and anti-glycolipid antibodies. *Brain.* 2002; 125: 2591-2625.

[4] Chiba A, Kusunoki S, Shimizu T, et al. Serum IgG antibody to gangalioside GQ1b is a possible marker of Miller Fisher syndrome. *Ann. Neurol.* 1992; 31: 677-679.

[5] Chiba A, Kusunoki S, Otaba H, et al. Serum anti-GQ1b IgG antibody is associated with ophthalmoplegia in Miller Fisher syndrome and Guillain-Barré syndrome: clinical and immunohistochemical studies. *Neurology.* 1993; 43: 1911-1917.

[6] Kusunoki S, Chiba A, Kanazawa I. Anti-GQ1b IgG antibody is associated with ataxia as well as ophthalmoplegia. *Muscle Nerve.* 1999; 22: 1071-1074.

[7] Yuki N, Sato S, Tsuji S, et al. An immunologic abnormality common to Bickerstaff's brainstem encephalitis and Fisher's syndrome. *J. Neurol. Sci.* 1993; 118: 83-87.

[8] Chiba A, Kusunoki S, Obata H, et al. Ganglioside composition of the human cranial nerves, with special reference to pathophysiology of Miller Fisher syndrome. *Brain Res.* 1997; 745: 32-36.

[9] Kuwabara S, Asahina M, Nakajima M, et al. Special sensory ataxia in Miller Fisher syndrome detected by postural body sway analysis. *Ann. Neurol.* 1999; 45: 533-536.
[10] Plomp JJ, Molenaar PC, O'Hanlon GM, et al. Miller Fisher anti-GQ1b antibodies; alpha-latrotoxin-like effects on motor end plates. *Ann. Neurol.* 1999; 45: 189-199.
[11] O'Hanlon GM, Plomp JJ, Chakrabarti M, et al. Anti-GQ1b ganglioside antibodies mediate complement-dependent destruction of the motor nerve terminal. *Brain.* 2001; 124: 893-906.
[12] Kaida K, Kusunoki S, Kanzaki M, et al. Anti-GQ1b antibody as a factor predictive of mechanical ventilation in Guillain-Barré syndrome. *Neurology.* 2004; 62: 821-824.
[13] Kuzumoto Y, Shioyama M, Kihara M, et al. Abnormal sudomotor axon reflex and antiganglioside antibodies. *Muscle Nerve.* 2006; 33: 828-829.
[14] Yuki N, Taki T, Takahashi M, et al. Molecular mimicry between GQ1b ganglioside and lipopolysaccharides of Campylobacter jejuni isolated from patients with Fisher's syndrome. *Ann. Neurol.* 1994; 36: 791-793.
[15] Jacobs BC, Endtz H, van der Meche FG, et al. Serum anti-GQ1b IgG antibodies recognize surface epitopes on Campylobacter jejuni from patients with Miller Fisher syndrome. *Ann. Neurol.* 1995; 37: 260-264.
[16] Kusunoki S, Chiba A, Kon K, et al. N-acetylgalactosaminyl GD1a is a target molecule for serum antibody in Guillain-Barré syndrome. *Ann. Neurol.* 1994; 35: 570-576.
[17] Yuki N, Yoshino H, Sato S, et al. Acute axonal polyneuropathy associated with anti-GM1 antibodies following Campylobacter enteritis. *Neurology.* 1990; 40: 1900-1902.
[18] van den Berg LH, Marrink J, de Jager AE, et al. Anti-GM1 antibodies in patients with Guillain-Barré syndrome. *J. Neurol. Neurosurg. Psychiatry.* 1992; 55: 8-11.
[19] Kuwabara S, Yuki N, Koga M, et al. IgG anti-GM1 antibody is associated with reversible conduction failure and axonal degeneration in Guillain-Barré syndrome. *Ann. Neurol.* 1998; 44: 202-208.
[20] Rees JH, Gregson NA, Hughes RAC. Anti-ganglioside GM1 antibodies in Guillain-Barré syndrome and their relationship to Campylobacter jejuni infection. *Ann. Neurol.* 1995; 38: 809-816.
[21] Corbo M Quattrini A, Latov N, et al. Localization of GM1 and Gal(beta1-3)GalNAc antigenic determinants in peripheral nerve. *Neurology.* 1993; 43: 809-814.
[22] Kusunoki S, Chiba A, Tai T, et al. Localization of GM1 and GD1b antigens in the human peripheral nervous system. *Muscle Nerve.* 1993; 16: 752-756.
[23] Yuki N, Yamada M, Koga M, et al. Animal model of axonal Guillain-Barré syndrome induced by sensitization with GM1 ganglioside. *Ann. Neurol.* 2001; 49: 712-720.
[24] Yuki N, Taki T, Inagaki F, et al. A bacterium lipopolysaccharide that elicits Guillain-Barré syndrome has a GM1 ganglioside-like structure. *J. Exp. Med.* 1993; 178: 1771-1775.
[25] Miyazaki T, Kusunoki S, Kaida K, et al. Guillain-Barré syndrome associated with IgG monospecific to ganglioside GD1b. *Neurology.* 2001; 56: 1227-1229.
[26] Kusunoki S, Shimizu J, Chiba A, et al. Experimental sensory neuropathy induced by sensitization with ganglioside GD1b. *Ann. Neurol.* 1996; 39: 424-431.
[27] Kaida K, Kusunoki S, Kamakura K, et al. Guillain-Barré syndrome with antibody to a ganglioside, N-acetylgalactosaminyl GD1a. *Brain.* 2000; 123: 116-124.

[28] Kaida K, Kusunoki S, Kamakura K, et al. GalNAc-GD1a in human peripheral nerve. Target sites of anti-ganglioside antibody. *Neurology.* 2003; 61: 465-470.

[29] Kaida K, Kusunoki S, Kamakura K, et al. Guillain-Barré syndrome with IgM antibody to the ganglioside GalNAc-GD1a. *J. Neuroimmunol,* 2001; 113:260-267.

[30] Saida T, Saida K, Dorfman SH, et al. Experimental allergic neuritis induced by sensitization with galactocerebroside. *Science.* 1979; 204: 1103-1106.

[31] Kusunoki S, Chiba A, Hitoshi S, et al. Anti-Gal-C antibody in autoimmune neuropathies subsequent to mycoplasma infection. *Muscle Nerve.* 1995; 18: 409-413.

[32] Kusunoki S, Shiina M, Kanazawa I. Anti-Gal-C antibodies in GBS subsequent to mycoplasma infection: Evidence of molecular mimicry. *Neurology.* 2001; 57: 736-738.

[33] Ho TW, Willison HJ, Nachamkin I, et al. Anti-GD1a antibody is associated with axonal but not demyelinating forms of Guillain-Barré syndrome. *Ann. Neurol.* 1999; 45: 168-173.

[34] Yako K, Kusunoki S, Kanazawa I. Serum antibody against a peripheral nerve myelin ganglioside, LM1, in Guillain-Barré syndrome. *J. Neurol. Sci.* 1999; 168: 85-89.

[35] Irie S, Saito T, Nakamura K, et al. Association of anti-GM2 antibodies in Guillain-Barré syndrome with acute cytomegalovirus infection. *J. Neuroimmunol.* 1996; 68: 19-26.

[36] Hakomori S. The glycosynapse. *Proc. Natl. Acad. Sci. USA.* 2002; 99: 225-232.

[37] Kaida K, Morita D, Kanzaki M, et al. Ganglioside complexes as new target antigens in Guillain-Barré syndrome. *Ann. Neurol.* 2004; 56: 567-571.

[38] Simons K, Ikonen E. Functional rafts in cell membranes. *Nature.* 1997; 387: 569-572.

[39] Kaida K, Kanzaki M, Morita D, et al. Anti-ganglioside complex antibodies in Miller Fisher syndrome. *J. Neurol. Neurosurg. Psychiatry.* 2006; 77: 1043-1046.

[40] Nobile-Orazio E, Manfredini E, Carpo M, et al. Frequency and clinical correlates of anti-neural IgM antibodies in neuropathy associated with IgM monoclonal gammopathy. *Ann. Neurol.* 1994; 36: 416-424.

[41] Ariga T, Kohriyama T, Freddo L, et al. Characterization of sulfated glucuronic acid containing glycolipids reacting with IgM M-proteins in patients with neuropathy. *J. Biol. Chem.* 1987; 262: 848-853.

[42] Tatum AH. Experimental paraprotein neuropathy, demyelination by passive transfer of human IgM anti-myelin-associated glycoprotein. *Ann. Neurol.* 1993; 33: 502-506.

[43] Shiina M, Kusunoki S, Miyazaki T, et al. Variability in immunohistochemistries of IgM M-proteins binding to sulfated glucuronyl paragloboside. *J. Neuroimmunol.* 2001; 116: 206-212.

[44] Willison HJ, O'Leary CP, Veitch J, et al. The clinical and laboratory features of chronic sensory ataxic neuropathy with anti-disialosyl IgM antibodies. *Brain.* 2001; 124: 1968-1977.

[45] Kusunoki S, Hitoshi S, Kaida K, et al. Degeneration of rabbit sensory neurons induced by passive transfer of anti-GD1b antiserum. *Neurosci. Lett.* 1999; 273: 33-36.

[46] Kusunoki S, Hitoshi S, Kaida K, et al. Monospecific anti-GD1b IgG is required to induce rabbit ataxic neuropathy. *Ann. Neurol.* 1999; 45: 400-403.

[47] Hitoshi S, Kusunoki S, Tsuji S, et al. Rabbit experimental sensory ataxic neuropathy; Anti-GD1b antibody-mediated trkC downregulation of dorsal root ganglia neurons. *Neurosci. Lett.* 1999; 260: 157-160.

In: Glycolipids: New Research
Editor: Daiki Sasaki, pp. 133-144
ISBN: 978-1-60456-216-3

Chapter VI

Domain Coexistence in Multi-Component Model Membranes

Dolores C. Carrer*
Group of Biophysics, Biotechnology Centre – TU Dresden,
01307 Dresden, Germany

Abstract

The membrane raft hypothesis proposed that membrane lipid domains enriched in Chol and sphingolipids form dynamic sorting and signalling platforms for membrane proteins in cells. Cell membranes however cannot be accurately described with only the two categories of raft or non-raft; instead we should talk about different domains which should be characterized each in all its complexities. In this chapter the current picture on membrane domains is discussed, and the usefulness of model systems to understand molecular configuration is exemplified by monolayer isotherms, calorimetry and spectroscopy studies in model membranes containing ceramide. These results are interpreted with the aim of explaining the ability of ceramide to produce very large domains or "platforms" in cell membranes.

Introduction

The membrane raft hypothesis proposed that membrane lipid domains enriched in Chol and sphingolipids form dynamic sorting and signalling platforms for membrane proteins in cells. The coexisting domains are thought to be liquid ordered/liquid disordered, with the liquid disordered phase being the major, percolating phase and the liquid ordered phase being a minor phase, represented by domains that would float like rafts in a sea of disordered phase. These isolated ordered domains would be able to locally enrich selected proteins and thus allow for more efficient interactions between, for example, the proteins that give rise to the

* E-mail: dolores.carrer@biotec.tu-dresden.de

different steps of a signalling event. Since propagation of a signal transduction event usually involves protein-protein interactions, such events are highly enhanced if the proteins involved are physically near, concentrated in the same membrane domain. This hypothesis has received enormous attention and brought basic membrane biophysics to the attention of biologists.

We Should Talk about Domains, Not Rafts

One wide-spread definition of rafts has relied on assessment of membrane heterogeneity by membrane lysis with non-ionic detergents, such as Triton X-100 at 4 °C. A second definition for lipid rafts has been cellular processes that are perturbed by either cholesterol or sphingomyelin depletion. In works where the sphingolipid under scrutiny has been ceramide, the term raft has been used to designate the ordered domains enriched in ceramide.

Works showing that membranes composed of lipids that are resistant to detergent solubilization are apparently in a liquid-ordered state have been the basis for proposing that lipid rafts in plasma membranes represent the liquid-ordered phase of coexisting liquid-ordered and liquid disordered phases [Edidin 2003]. In the case of membranes containing ceramide, it is often not clear whether the ordered domains are in a gel state or in a liquid ordered state or in some other kind of physical state not yet characterized, even though the word "raft" implies they are in a liquid ordered state and in doing so allows the reader to believe the physical state of these domains is already completely understood [Megha et al. 2006, Bollinger et al. 2005]. Also, it should be taken into account that protein interaction with membranes can induce the formation of domains that would not necessarily be in a different phase state, but could be different as regards to headgroup structure [Hinderliter et al. 2004]. This last kind of domain is probably ubiquitous in the cell plasma membrane. However, when domains in the membrane are classified as either "rafts" or "non rafts", the possibility of variety inside the categories "raft" or "non raft" is neglected.

In this chapter I will concentrate in particular on domains enriched in ceramide and I will purposefully avoid calling them "rafts", even though ceramide-enriched domains are commonly classified in the literature as belonging to the raft domains. This I think can help us approach the study of these particular membrane structures with a more unbiased understanding.

Ceramide has been observed to be involved in the initiation of apoptosis, in bacterial infection and in rhinovirus infection. It has been observed in cells to lead to the production of large domains or "platforms" and protein rearrangements, most notably the capping of the death receptor CD95/Fas [Bollinger et al. 2005].

Domains in Cells

Experiments performed in cells suggest that many proteins partition into sphingolipid-cholesterol domains, at least under some conditions: glycosylphosphatidylinositol (GPI)-anchored proteins, G-proteins, some transmembrane proteins, some transporters. However, when studied in model membranes, protein partition into coexisting liquid ordered and liquid disordered phases favour the liquid disordered phase in most of the cases. One notable exception is the observation that GPI-anchored PlAP not only partitions preferentially into liquid ordered domains in model membranes, but when a long chain ceramide is present in the membrane, forming ceramide-enriched gel phase domains, the GPI-anchored PlAP partitions preferentially to these gel phase domains over either the liquid ordered or the liquid disordered phase [Chiantia et al. 2007].

On the other hand, consensus data on the size, lifetime, and even the existence of lipid domains in intact plasma membranes have been elusive. Estimates of lipid domain sizes in the literature range form several micrometers to just a few molecules [Hao et al. 2001, Mayor and Rao 2004]. Although scarce, there is some direct evidence to the presence of liquid-liquid domain coexistence in the plasma membrane of intact cells at 37 °C. One such work is Meder et al. 2005. In this work the authors find that the plasma membrane of two lines of polarized epithelial cells shows liquid-liquid domain coexistence, with the liquid-ordered phase being the percolating phase at temperatures below 25 °C and the minor, disconnected phase at 37°C. Also Gaus and co authors find micrometer-sized domains with Laurdan GP values corresponding to a liquid ordered phase on live cells at 37 °C [Gaus et al. 2003] . These domains represent at most 15% of the surface of the cells at 37 °C, with the surface coverage increasing to 50% at lower temperatures. Another kind of lipid domain that has been observed in the plasma membrane of living cells is ceramide-enriched membrane domains. Several studies have demonstrated by fluorescence microscopy the binding of two different anti-ceramide antibodies to the plasma membrane of cells in which the ceramide signalling pathway had been activated [Bollinger et al. 2005]. These domains have been shown to be very large, possibly formed by the coalescence of smaller domains into large "platforms". The physical state of ceramide-enriched domains is however not studied in works done in cells. Also in model systems, the physical state of ceramide-enriched domains in complex mixtures is still not well understood.

While the study of membrane organization in the plasma membrane of intact, living cells is the most relevant to cell biology, the enormous complexity of cellular plasma membrane protein and lipid compositions precludes the complete understanding of the molecular mechanisms involved both in lipid domain formation and in the interaction of these domains with membrane proteins. In order to better understand the physicho-chemical processes involved, model systems are the ideal tool.

What Model Systems Tell about Domains

To understand the plasma membrane, scientists draw their resources from the physical chemistry of model membranes, that is membranes that self assemble in vitro from a limited and well-known amount of proteins and lipids or membranes made of a very complex mixture that is taken from natural membranes but put into a spatial arrangement that is easy to study, such as a monolayer at the air-water interface or a giant unilamellar vesicle for example [Oliveira and Maggio 2002, Bernardino de la Serna et al. 2004].

Model membrane systems have been studied for more than a century. Membrane physical chemistry started with observations of the spreading of oils on water (for an excellent review on the history of the pioneers of membrane chemistry and physics, see Tanford 2004). Monolayer experiments at the air-water interface where the basis that Gorter and Grendel used to infer that erythrocyte membranes are in fact lipid bilayers [Gortel and Grendel 1925]. Since then the picture has gotten increasingly complex and more detailed, including not only the spatial arrangement of proteins and lipids in cell membranes but also the dynamics of this arrangement. Phase separation in phospholipid bilayers was described in the 1970s [Schimshick and McConnell 1973a, Mabrey and Sturtevant 1976]. These studies dealt mainly with separation into coexisting gel and liquid disordered phases. Other studies dealt with mixtures of cholesterol and phospholipids; in this case liquid / liquid phase coexistence was described, either from experimental data or from theoretical models [Schimshick and McConnell 1973b, Snyder and Freire 1980, Ipsen et al. 1987]. A phase diagram was described where a liquid ordered phase and a liquid disordered phase form and coexist with each other in the plane of the membrane in a certain range of compositions and temperatures [Ipsen et al. 1987]. The liquid ordered phase is very peculiar, with characteristics intermediate between a gel phase and a fluid phase: it displays fast lateral diffusion of the lipids but these maintain a high degree of order in the acyl chains. The liquid disordered phase shows fast lateral diffusion but highly disordered acyl chains, while the lipids in a gel phase have very low lateral diffusion and highly ordered acyl chains. Liquid ordered and liquid disordered phases have also been seen to coexist with each other in mixtures of lipids that mimic the outer hemilayer of the plasma membrane: one lipid with a low melting temperature, one lipid with a high melting temperature and cholesterol. This kind of ternary lipid mixture has become a canonical model system to mimic the plasma membrane.

Using giant plasma membrane vesicles, derived form cells by chemically induced plasma membrane vesiculation, Baumgart and coauthors have observed micrometer scale coexisting fluid phases [Baumgart et al. 2007]. Since these membranes contain all the components of the cell plasma membrane except for the interacting cytoskeleton, it is proven that an increase in complexity does not preclude liquid/liquid coexistence nor does it bring about a different qualitative behaviour when compared to model membranes. In most studies however, the plasma membrane of cells only displays, by confocal microscopy, liquid/liquid domain coexistence at low temperatures (10-23 °C). At 37 °C, most studies (with the notable exception of studies on ceramide-enriched domains) conclude that if there are domains present they must be smaller than 200 nm, so that they can not be detected by light microscopy. The conclusion would be then that, at physiological temperatures in the plasma membrane of most cells, the membrane would only show transient, small-scale heterogeneities.

To understand this, it helps to go back to thermodynamics of model membrane systems. By measuring domain formation in simple lipid mixtures, it is concluded that lipid-lipid interaction energies are quite small [Dewa et al. 1996, Sugahara et al. 2003]. If lipid-lipid interactions were large, formation of lipid domains would be irreversible, leading to static structures. This would obviously not be consistent with the normal function of a cell plasma membrane, which is constantly being renewed by endo- and exocytosis, lipid synthesis and degradation, receptor internalization, protein turnover, etc, and so it is an extremely dynamic and plastic structure. Indeed, when measured in model systems, interaction Gibbs energies between lipid species vary between -200 and +200 cal/ml, and only seldom reach the +400 cal/mol required for complete phase separation [Almeida et al. 2005]. When the number of molecules involved is very big, these interactions are amplified leading to domain formation, but the process is reversible because the energy barriers are small. The study of protein-membrane interaction in model systems shows also that for example peripheral membrane protein interaction with a model membrane can induce the formation of domains in a reversible way [Hinderliter et al. 2004]. It must be noted that domain formation induced by peripheral membrane proteins does not necessarily produce phase separation, but can induce only compositional domain separation if the membrane-lipid interaction is mediated primarily by the nature of the lipid headgroup. If, however, one particular kind of headgroup is present in molecules with mostly saturated or very long acyl chains, then it is possible that the electrostatic interaction of the peripheral membrane protein will produce not only compositional domain formation but also the separation of the membrane into domains that are in different phase states.

At physiological temperatures then, it is most likely that the membrane would only show short-lived, small-scale heterogeneities, which could be stabilized to more long-lived, meso-scale domain coexistence only by an important change in either the local lipid composition or the local concentration/conformation of proteins.

Phase Diagrams

As a further example of what can be learned from simple model systems, I will discuss the case of ceramide. Short chain ceramides are frequently used to mimic the effect of rising ceramide concentration in cells undergoing the first steps of apoptosis. Figure 1a shows the calorimetry curves obtained from mixtures of N-octanoyl sphingosine (C8Cer) with dimiristoylphosphatidilcholine (DMPC). From these curves, a phase diagram can be constructed by determining the temperature at which the transition starts (T_S) and the temperature at which the transition ends (T_L). Figure 1b shows how T_S and T_L are measured from the calorimetry data. These temperatures are then plotted as a function of composition to construct a phase diagram as shown in Figure 2. In a diagram such as shown in Figure 2, a horizontal line in T_S indicates that below that temperature, two different gel phases coexist. This can easily be understood by realizing that, if when the concentration of C8Cer rises the melting of the bilayer always starts at the same temperature it is because the calorimeter is detecting the heat uptake of a phase where no "foreign" molecules are being incorporated, and so it undergoes its transition always at the same temperature with no regard for the amount of the second component. This phase is composed of the first component, the second component must then be partitioning to a different (gel) phase. When the C8Cer concentration reaches a value of ≈30 mole% however, the T_S changes to a new, higher value.

The fact that the width of the peak at this composition is very narrow indicates the melting of a stoichiometric compound, a mixture that behaves as a pure component. From the phase diagram in Figure 2, it is possible to know the amount and the composition of each coexisting phase by using the lever rule. In point a for example, half of the molecules in the bilayer are forming a (almost) pure DMPC phase, while the other half are participating in the formation of a phase that has a composition DMPC:C8Cer 7:3.

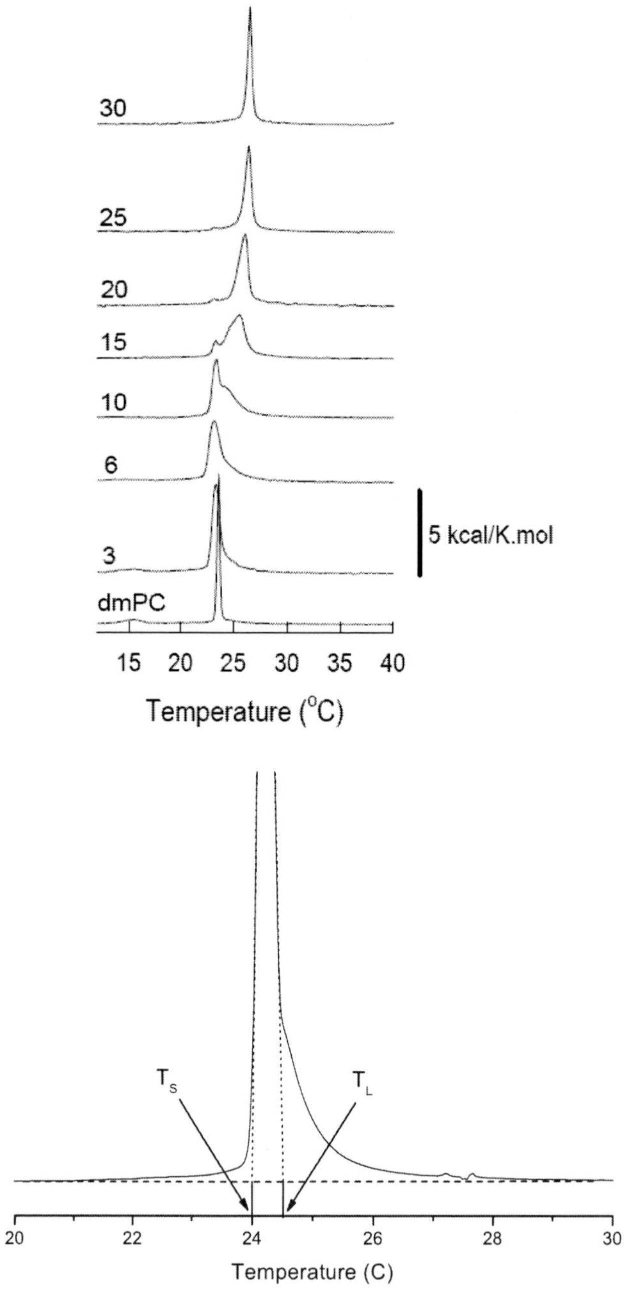

Figure 1. Calorimetric curves of DMPC/C8Cer mixtures. (a) The heat capacity as a function of temperature is shown for DMPC/C8Cer mixtures. The figures in the curves indicate, in mole%, the amount of C8Cer in the mixtures. The ordinate values for DMPC are multiplied by 0,5. Reprinted with permission from Biophys. J. 90: 2394-2403. Copyright (2006) Biophysical Society. (b) An example of how T_S and T_L are measured from a calorimetric curve.

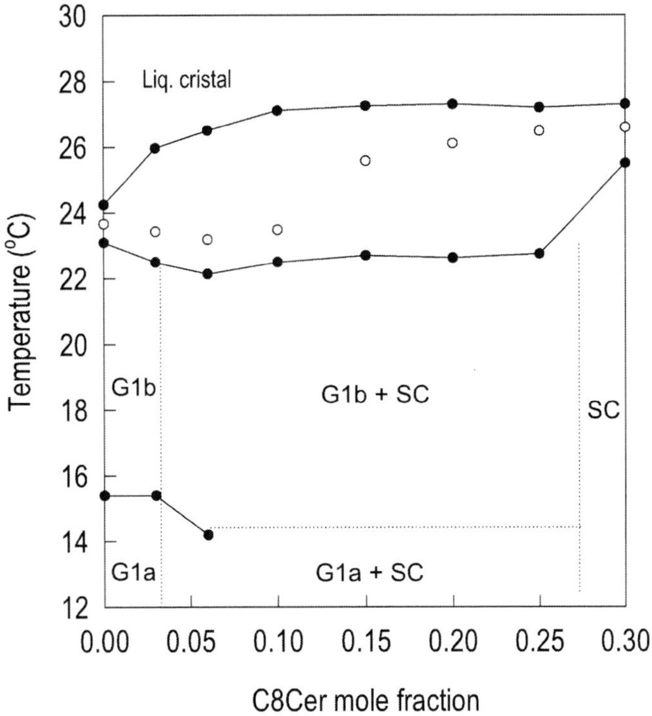

Figure 2. Partial calorimetric phase diagram of DMPC/C8Cer mixtures. Calorimetrically determined Tp, T_S, T_L (solid symbols) and Tm (open symbols) are shown. Reprinted with permission from Biophys. J. 90: 2394-2403. Copyright (2006) Biophysical Society.

An interesting observation drawn from calorimetry of phospholipid-ceramide mixtures is resumed in Table 1.

Table 1. Normalized transition entropy for ceramide-containing systems. The transition entropy $\Delta S = (\Delta H/T_m)$ is normalized by the expected transition entropy ΔS_{exp} = (1.9 e.u./mol (n1+n2)) (Huang and Mason 1986)

X_{C8Cer}*	T_m (°C)	ΔH (kcal/mol)	$\Delta S/\Delta S_{exp}$	X_{BBCer}[#]	T_m (°C)	ΔH (kcal/mol)	$\Delta S/\Delta S_{exp}$
0 (dmPC)	23.67	7.4	0.47	0 (dpPC)	40.93	8.4	0.44
0.06	23.19	6.9	0.44	0.04	41.77	8.1	0.42
0.15	25.57	6.3	0.41	0.16	42.43	7.4	0.38
0.25	26.48	5.8	0.38	0.20	43.07	6.5	0.33
0.30	26.59	4.9	0.32	0.30	52.07	6.1	0.30

*Mol fraction of C8Cer; [#] mol fraction of bovine brain ceramide.

Reprinted with permission from Biophys. J. 90: 2394-2403. Copyright (2006) Biophysical Society.

In this table, that summarizes calorimetry data from ceramide-pospholipid mixtures, it can be seen that increasing concentrations of either natural mixtures of ceramides (where 30% of the molecules have the fatty acyl chain much longer than the sphingosine chain) or asymmetric ceramides (with the fatty acyl chain much shorter that the sphingosine chain) produce a decreasing value of the phase transition entropy change. This could be explained by either an increased order of the fluid phase, an increased disorder of the gel phase, or both effects simultaneously. To know which is the case, we need to use techniques that are more sensitive to local molecular configuration.

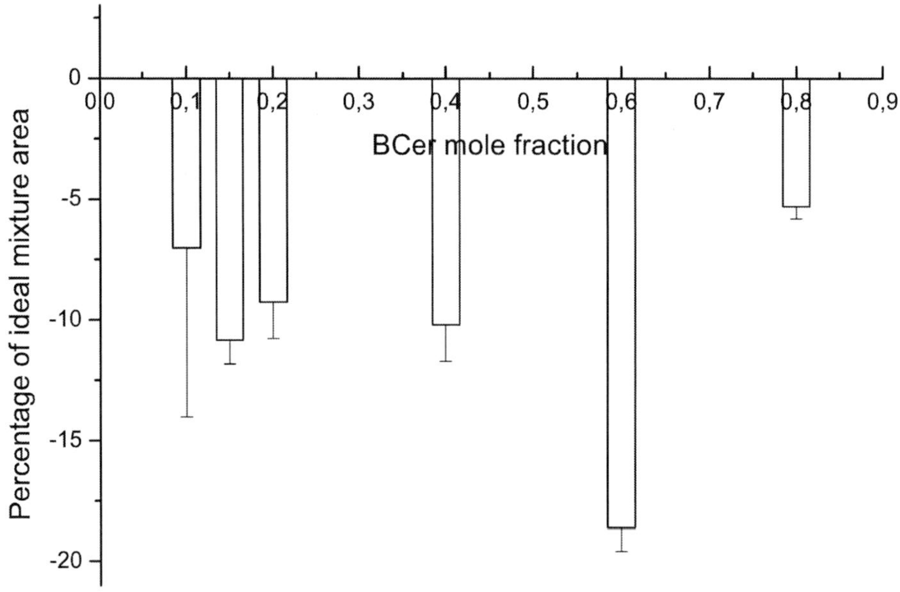

Figure 3. Molecular area in mixed monolayers of DPPC and brain ceramides. Variation of mean molecular area with respect to the molecular area of an ideal mixture is shown as a function of brain ceramide concentration in the monolayer, at 10 mN/m and 27°C.

Getting into Details

A phase diagram can tell us a lot about the macroscopic behaviour of the sample, but to know details at the molecular level different techniques are needed. Experimental data show a dual effect of ceramide on bilayers depending on whether the membrane is in the fluid or in the gel phase. H^2-RMN performed on mixtures of DPPC and brain ceramides that contain deuterated phospholipid shows that the order parameters of the PC are incremented by the presence of ceramide at high temperatures, indicating an enhanced interaction between the acyl chains [Huang et al. 1996]. A condensing effect of brain ceramides in the liquid expanded state (analog to the fluid phase in bilayers) is observed in mixed dipalmitoylphosphatidylcholine – brain ceramide monolayers [Carrer and Maggio 1999]. Figure 3 shows, at 10 mN/m, the percentage of molecular area reduction with respect to the ideal mixture as a function of ceramide concentration in dipalmitoylphosphatidylcholine – brain ceramide mixtures. Brain ceramides produce a packing of the molecules in the monolayer that is more dense than what is to be expected for an ideal mixture, in which molecules behave equally when they are mixed with like or unlike molecules. This increased packing (of up to 20% of molecular area) is however only observed at pressures below the phase transition pressure of these mixtures, that is in the liquid expanded state of the monolayer, analogous to the liquid disordered state of a bilayer. Also in mixtures of synthetic ceramides with DPPC it has been shown by RMN that ceramides increase by 10% the order parameter and produce a decrease in superficial area per molecule of DPPC in the fluid phase [Huang et al. 1998]. In mixtures of DMPC with natural mixtures of ceramide, fluorescence anisotropy shows that ceramide diminishes the free volume and increases packing in the fluid phase, while a disordering effect is seen in the gel phase [Holopainen et al. 1997]. Increased microviscosity in the fluid phase is also seen in mixtures of DOPC and C16Cer [Holopainen et al. 1998]. In mixtures of DMPC with C8Cer, the incorporation of a small amount of spin-labeled fatty acids allows for the measurement of local order at different planes in the bilayer by EPR. We showed in this study that C8Cer induces an increase in acyl chain order in the fluid phase, while producing a decrease in acyl chain order in the gel phase, especially in the deeper parts of the bilayer away from the water-hydrocarbon interface [Carrer et al. 2006].

This dual effect of asymmetric ceramides, that produce an increased order in the fluid phase while disordering the gel phase, can be understood by observing that ceramide has a very small head group, which induces condensation of the phospholipid molecules, increasing packing and order especially in the vicinity of the interface. In the deeper parts of the bilayer however, the asymmetry of the chains produce and abundance of bulky methyl groups not only in the middle plane of the bilayer, but also at other heights in the bilayer, in positions where they disrupt packing. Asymmetric ceramides introduce in the bilayer also acyl chains that are either too short or too long to pack appropriately with the neighbouring chains. The ordering effect near the interface produces increased order both in the fluid and in the gel phase, with a more marked effect in the fluid phase since the gel phase is already in a very ordered configuration. The disordering effect in the depths of the membrane has a much bigger effect in the gel phase than in the fluid phase, where the chains are already very disordered. The result of these combined effects would be then the production of a disordered gel phase and an ordered fluid phase, making both more similar to each other and diminishing the entropy change of the transition. This is very similar to the effect that cholesterol has in membranes, where it produces a new phase –the liquid ordered phase- that has characteristics intermediate between those of a liquid and a gel phase.

To summarize, spectroscopy and calorimetry studies on mixtures of ceramides and phospholipids show that ceramides produce an increment in the chain-chain interactions, increasing order and microviscosity in the fluid phase but doing the contrary in the gel phase. The observed decrease in the entropy change of the phase transition of mixtures of asymmetric ceramides can then be explained by a dual effect of ceramides that is remarkably similar to the effect that cholesterol has in phospholipid membranes.

Ceramides and Cholesterol

Other authors have put forward evidence that ceramide is able to displace cholesterol from ordered lipids domains [Megha and London 2004, Yu et al. 2005]. Using lipid vesicles containing co-existing liquid ordered and liquid disordered phases, they have found that ceramides get enriched in detergent-resistant membranes at the expense of cholesterol.

Ceramide also displaces dehidroergosterol from the ordered to the disordered phase as assessed by fluorescence quenching assays [Megha and London 2004]. In mixtures of DOPC, stearoyl sphingomyelin (C18SM), N-stearoyl sphingosine (C18Cer) and cholesterol, we have directly visualized for the first time cholesterol displacement from ordered domains: in these mixtures, ceramide forms gel-phase domains that exclude the fluorescent cholesterol analog Bodipy-Cholesterol2 [Chiantia et al. 2008]. The effect of a natural mixture of ceramides, with a complex mixture of fatty acids, is however different from the effect of C18Cer. In giant unilamellar vesicles, we have observed that the domains formed in mixtures containing cholesterol and brain ceramides along with unsaturated phospholipid are not round but linear or flower-shaped, pointing to a possible decrease in line tension between the coexisting domains. Preliminary data on the diffusion coefficient in these coexisting phases seems to indicate that the liquid disordered phase is getting enriched in cholesterol while the liquid ordered phase does not seem affected. Based on these observations, and on the work discusser earlier that shows how the effect of natural cermides in membranes is very similar to that of cholesterol, I would like then to take the description of cholesterol-ceramide interaction one step further: ceramide seems not only to be able to displace but also to replace cholesterol in ordered domains, especially if it has asymmetric chains. This ability of ceramide would multiply its effects on membranes because the total amount of membrane affected would be not only the small region in which ceramide is produced, but also the surrounding region to where cholesterol is displaced. This could be a mechanism to explain the observation, done in cells, that the increase in ceramide concentration in the initial steps of apoptosis produces very big "platforms" or caps, affecting protein dynamics and spatial distribution.

Conclusion

Cell membranes are extremely complex and dynamic structures that cannot be accurately described with only the two categories of rafts or non- rafts. This simplistic view distorts the understanding of membrane organization; detecting and characterizing different classes of membrane domains is very much needed. Other scientists have started this process; see for example Shaikh and Edidin 2006 and Almeida et al. 2005. I hope I have taken some steps towards this goal also in this chapter.

References

Almeida P. F. F., Pokorny A., Hinderliter A. (2005) Thermodynamics of membrane domains. *Biochim. Biophys. Acta, 1720,* 1-13.

Baumgart T., Hammond A. T., Sengupta P., Hess S. T., Holowka D. A., Baird B. A., Webb W. W. (2007) Large-scale fluid/fluid phase separation of proteins and lipids in giant plasma membrane vesicles. *Proc. Natl. Acad. Sci. U.S.A., 104*, 3165-3170.

Bernardino de la Serna J., Perez-Gil J., Simonsen A. C., Bagatolli L. A. (2004) Colesterol rules: direct observation of the coexistence of two fluid phases in native pulmonary surfactant membranes at physiological temperatures. *J. Biol. Chem., 39,* 40715-722.

Bollinger C. R., Teichgräber V., Gulbins E. (2005) Ceramide-enriched membrane domains. *Biochim. Biophys. Acta, 1746*, 284-294.

Carrer D. C., Maggio B. (1999) Phase behaviour and molecular interactions in mixtures of ceramide with dipalmitoylphosphatidylcholine. *J. Lipid Res., 40,* 1978-1989.

Carrer D. C., Schreier S., Patrito M., Maggio B. (2006) Effects of a short-chain ceramide on bilayer domain formation, thickness and chain mobility: DMPC and asymmetric ceramide mixtures. *Biophys. J., 90,* 2394-2403.

Chiantia S., Ries J., Chwastek G., Carrer D., Li Z., Bittman R., Schwille P. (2008) Role of ceramide in membrane protein organization investigated by simultaneous AFM and FCS. Submitted to *Biochim. Biophys. Acta.*.

Dewa T., Vigmond S. J., Regen S. L. (1996) Lateral heterogeneity in fluid bilayers compsed of saturated and unsaturated phospholipids. *J. Am. Chem. Soc., 118*, 3435-3440.

Edidin M. (2003) The state of lipid rafts: from model membranes to cells. *Ann. Rev. Biophys. Biomol. Struct., 32*, 257-283.

Gaus K., Gratton E., Kable E. P. W., Jones A. S., Gelissen I., Kritharides L., Jessup W. (2003) Visualizing lipid structure and raft domains in living cells with two-photon microscopy. *Proc. Natl. Acad. Sci. USA,100*, 15554-15559.

Gorter E. and Grendel F. (1925) On bimolecular layers of lipoids on the chromocytes of the blood. *J. Exp. Med., 41*, 439-443.

Hao M. M., Mukherjee S., Maxfield F. R. (2001) Cholesterol depletion induces large scale domain segregation in living cell membranes. *Proc. Natl. Acad. Sci U.S.A., 98*, 13072-077.

Hinderliter A., Biltonen R. L., Almeida P. F. F. (2004) Lipid modulation of protein-induced membrane domains as a mechanism for controlling signal transduction. *Biochemistry, 43,* 7102-7110.

Holopainen J. M., Lehtonen J. Y. A., Kinnunen P. K. J. (1997) Lipid microdomains in dimyristoylphosphatidylcholine-ceramide liposomes. *Chem. Phys.Lipids, 88*, 1-13.

Holopainen J. M., Subramanian M., Kinnunen P. K. J. (1998) Sphingomyelinase induces lipid microdomain formation in a fluid phosphatidylcholine/sphingomyelin membrane. *Biochemistry, 37*, 17562-70.

Huang H., Goldberg E. M., Zidovetzki R. (1996) Ceramide induces structural defects into phosphatidylcholine bilayers and activates phospholipase A_2. *Biochim.Biophys .Res. Com., 220*, 834-838.

Huang H., Goldberg E. M., Zidovetzki R. (1998) Ceramides perturb the structure of phosphatidylcholine bilayers and modulate the activity of phospholipase A2. *Eur. Biophys. J., 27*, 361-66.

Ipsen J. H., Karlstroem G., Mouritsen O. G., Wennerstroem H., Zuckermann M. J. (1987) Phase equilibria in the phosphatidylcholine-cholesterol system. *Biochim. Biophys. Acta, 905*: 162-172.

Mabrey S., Sturtevant J. (1976) Investigation of phase transitions of lipids and lipid mixtures by high sensitivity differential scanning calorimetry. *Proc. Natl. Acad. Sci. USA., 73,* 3862-866.

Mayor S., Rao M (2004) Rafts: scale-dependent, active lipid organization at the cell surface. *Traffic, 5,* 231-240.

Meder D., Moreno M. J., Verkade P., Vaz W. L. C., Simons K. (2005) Phase coexistence and connectivity in the apical membrane of polarized epithelial cells. *Proc. Natl. Acad. Sci. USA,103*, 329-334.

Megha, London E. (2004) Ceramide selectively displaces cholesterol from ordered lipid domains (rafts). *J. Biol. Chem., 279,* 9997-10004.

Megha, Bakht O., London E. (2006) Cholesterol precursors stabilize ordinary and ceramide-rich ordered lipid domains (lipid rafts) to different degrees. *J. Biol. Chem., 281*, 21903-21913.

Oliveira R. G., Maggio B. (2002) Compositional domain immiscibility in whole myelin monolayers at the air-water interface and Langmuir-Blodgett films. *Biochim. Biophys. Acta, 1561,* 238-250.

Tanford, C. (2004) *Ben Franklin stilled the waves: An informal history of pouring oil in water with reflections on the ups and downs of scientific life in general.* New York, Oxford University Press.

Shaikh S. R., Edidin M. A. (2006) Membranes are not just rafts. *Chem. Phys. Lipids, 144,* 1-3.

Schimshick E. J., McConnell H. M. (1973a) Lateral phase separation in phospholipid membranes. *Biochemistry, 12,* 2351-360.

Schimshick E. J., McConnell H. M. (1973b) Lateral phase separation in binary mixtures of cholesterol and phospholipids. *Biochem. Biophys. Res. Commun., 53,* 446-451.

Snyder B., Freire E. (1980) Compositional domain structure in phosphatidylcholine-cholesterol and sphingomyelin-cholesterol bilayers. *Proc. Natl. Acad. Sci. U.S.A., 77,* 4055-4059.

Sugahara M., Uragami M., Regen S. L. (2003) Selective association of cholesterol with long-chain phospholipids in liquid-ordered bilayers: support for the existence of lipid rafts. *J. Am. Chem. Soc., 125*, 13040-13041.

Yu C., Alterman M., Dobrowsky R. T. (2005) Ceramide displaces cholesterol from lipid rafts and decreases the association of the cholesterol binding protein caveolin-1. *J. Lipid Res., 46*, 1678-1691.

In: Glycolipids: New Research
Editor: Daiki Sasaki, pp. 145-156
ISBN: 978-1-60456-216-3

Chapter VII

Synthetic Ceramide Analogs for Anticancer Agents

Hyun-Joon Ha[a], Won Koo Lee[b] and Jungchan Park[c]
[a]Department of Chemistry and Protein Research center for Bio-Industry, Hankuk University of Foreign Studies, Yongin, 449-791, Korea
[b]Department of Chemistry and Interdisciplinary Program of Integrated Biotechnology, Sogang University, Seoul, 121-742, Korea
[c]Department of Bioscience and Biotechnology, and Protein Research center for Bio-Industry, Hankuk University of Foreign Studies, Yongin, 449-791, Korea

Abstract

Ceramide is an apoptosis-inducing signaling molecule in the sphingolipid metabolism which emerges as a potential target for anticancer therapy. Natural ceramide in solution has a uniquely folded butterfly conformer with strong internal hydrogen bonding between amide and two alcohols at C1 and C3 and with hydrophobic interaction between two long alkyl chain of sphingosine backbone and fatty acid. Its heterocyclic analogs as 3-alkanoyl or benzoyl-4-(1-hydroxyhexadec-2-enyl)-oxazolidin-2-ones were designed by binding of primary alcohol and amide in sphinogosine backbone as a carbamate. Those compounds without free primary alcohol have advantages not to perturb the ceramide metabolism most of which involved in the primary alcohol of sphingosine backbone. This structurally constrained analog of ceramide may have better activity compared with the natural ceramide without pre-organization of substrate, if necessary. They were synthesized by addition of acyl halide to the common ring, 4-(1-*t*-butyldimethylsilyloxyhexadec-2-enyl)-oxazolidin-2-one, which was elaborated from chiral aziridine-2-carboxylate including stereoselective reduction and ring opening reactions. All compounds were tested as antileukemic drugs against human leukemia HL-60 cells. Many of them including propionyl, cyclopentanoyl and *p*-nitrobenzoyl-4-(1-hydroxyhexadec-2-enyl)-oxazolidin-2-ones showed better antileukemic activities than natural C2-ceramide with good correlation between cell death and DNA fragmentation.

Cytotoxicity was induced by caspase activation without significant accumulation of endogenous ceramide concentration or any perturbation of ceramide metabolism.

Keywords: *ceramide analogs; oxazolidin-2-one, antileukemic activity; apoptosis.*

Sphingolipids and glycosphingolipids including ceramide, sphingomyelin, cerebroside, and ganglioside are unique membrane components of all eukaryotic cells (Kolter and Sandhoff, 1996; Dickson, 1998; Kolter and Sandhoff, 1999; Brodeser et al., 2003; Liao, et al. 2005). They play important roles in the regulation of cell proliferation, differentiation, migration, survival and cell-death via closely related metabolic pathways (Pettus et al., 2002; Brodesser, et al., 2003; Radin, 2003; Reynold et al., 2004). Among all metabolites ceramide is one of the most important molecules as a secondary messenger of sphingolipid signaling pathway (Reynold et al., 2004).

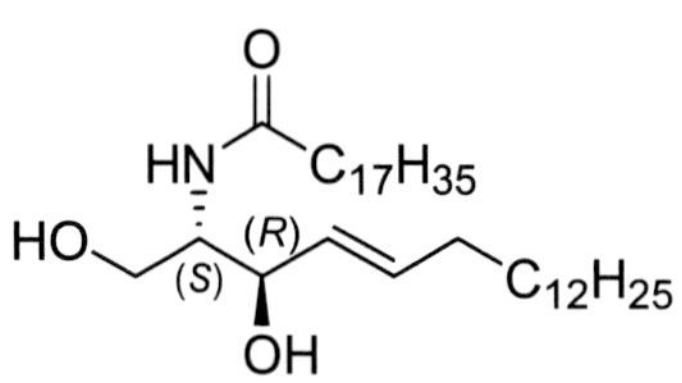

Ceramide (**1**)

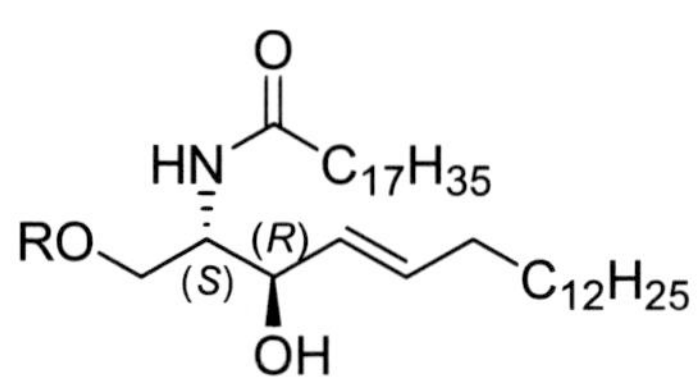

2 R = P(=O)$(O^-)_2$
3 R = P(=O)$(O^-)(OCH_2CH_2N^+Me_3)$
4 R = Glucose

The main structural components of ceramide (1) are a long alkyl chain with 2-amino-1,3-diols and C4,5-*trans*-olefin as the sphingosine backbone and an *N*-acyl group. In mammalian tissue the most common sphingoid base is *D-erythro*-C18-sphingosine [(2*S*,3*R*,4*E*)-2-amino-octadec-4-en-1,3-diol, (*E*)-sphing-4-enine] and the typical acyl group attached on the amine has the chain length of 16 - 24. The structural changes of ceramide at the primary alcohol occur at the cellular metabolic system by addition and removal of phosphate, phosphocholine and carbohydrates such as ceramide-1-phospahate (2), sphingomyelin (3) and glucosylceramide (4). Specifically ceramide is generated from sphingosine by ceramide synthase with an additional fatty acid chain to the sphingoid backbone. Hydrolysis of sphingomyelin bearing additional phosphocholine to ceramide is also important pathway for ceramide (Hannun, 1994; Hannun, 1996; Sawai and Hannun, 1999) as a secondary messenger by sphingomyelinase (Hannun et al., 2001) in response to extracellular agents and stresses including chemotherapeutic agents (Cifone et al., 1994), tumor necrosis factor α (TNF-α) (Kim et al., 1991), and ionizing radiation (Haimovitz-Friedman et al., 1994). The dynamic balance of the sphingolipid metabolites is an important factor whether a cell undergoes proliferation or programmed death so called apoptosis (Kinloch et al., 1999). The close relationship is found between ceramide and its hydrolytic products sphingosine triggered by growth factors such as platelet-driven growth factors. Ceramide as an important stimulating factor of apoptosis is hydrolyzed by ceramidase to sphingosine followed by activation of kinase to yield sphingosine-1-phosphate with cell growth.

Lots of sphingosine analogs were isolated from natural sources and synthesized based on the structural modification of sphingoid backbones, and their biological activities were evaluated. One of the man-made analogs Safingol and marine natural product spisulosin were good examples. Safingol is an inhibitor of protein kinase C (Rajewski et al., 1995; Sachs et al., 1995) and spisulosin inhibits the activity of Rho protein (Cuadros et al, 2000). Few years ago we envisaged the synthetic method for the construction of sphingoid backbone and their analogs including anticancer agent spisulosin and safingol based on the stereospecific reduction of 2-acylaziridine and aziridine ring opening reactions (Yun et al., 2003). These synthetic methods allow us to synthesize several new sphingosine analogs by modifying the long alkyl chain backbone, primary and secondary alcohols and amine with different configurations. Our synthetic sphingoid analogs were evaluated and identified as specific inhibitors of two mammalian sphingosine kinases 1 and 2. Sphingosine kinase 1 stimulates cell growth and survival but sphingosine kinase 2 induces apoptosis and suppression of cellular growth in various types (Liu et al., 2003). Among our synthetic analogs SG-12 and SG-14 had specific inhibitory effects selectively on human sphingosine kinase 2 while a well-known sphingosine kinase inhibitor *N,N'*-dimethylsphingosine (DMS) displayed inhibitory effect on both kinases (Kim et al, 2005). Most of active compounds retained potent cytotoxic effects comparable to *N,N'*-dimethylsphingosine. It is noteworthy that the phosphoric acid of SG-12 (DS-SG-44) was an agonist of sphingosine 1-phosphate (S1P). The S1P and DS-SG-44 induced sharp changes which were neither reproduced with $S1P_1/S1P_3$ receptor agonist VPC24191 or inhibited by the $S1P_1/S1P_3$ receptor antagonist VPC23019 (Kim et al. 2007).

Among the enormous increase in the research of sphingolipid metabolism the central metabolite ceramide attracts most attentions due to its important role at the stages of apoptosis with a selective toxicity to malignant but not to normal cells (Wang et al., 1999; Burow et al., 1998). Various analytical methods show that ceramide is accumulated during the initiation stage of apoptosis. Furthermore exogenous ceramide actually induces apoptotic programmed cell death. In particular, cell-permeable ceramide with short acyl group such as C2-Cer (*N*-Acetyl-*D-erythro*-C18-sphingosine) and C8-Cer (*N*-Octyl-*D-erythro*-C18-sphingosine) inhibit tumor cell growth and induce apoptosis in leukemia cell lines (Selzner et al., 2001; Fillet, et al., 2003). The ceramides with longer alkyl chain than C8-Cer including natural caramide (C18-Cer) are too lipophilic to pass through the cell membrane. Dihydroceramide (DHCer) with saturated *D-erythro*-C18-sphinganine does not have any biological activity which seems to be due to the lack of essential *trans* allylic alcohol (Karasavvas, et al., 1996; Javis, et al., 1996; Chang, Y. T. et al., 2002). For this reason, many synthetic and natural ceramide analogs were prepared and evaluated as anticancer agents by altering the backbone and by changing two hydroxy and one amine functional groups (Chart 1). That includes ceramide analogs whose backbone was modified with various aromatic or heteroaromatic systems (A, B and C) instead of simple long alkyl chain of sphingoid base (Chun et al, 2000; Bielawska et al., 1992; Bielawska et al., 1996). Especially, two compounds B and C known as D-MAPP and B13 (Bielawska et al., 1992; Bielawska et al., 1996) are able to block the conversion of ceramide to sphingosine followed by increase the endogenous ceramide level to induce apoptosis. The compound B13 prevented the growth of colon cancer effectively (Selzner et al., 2001).

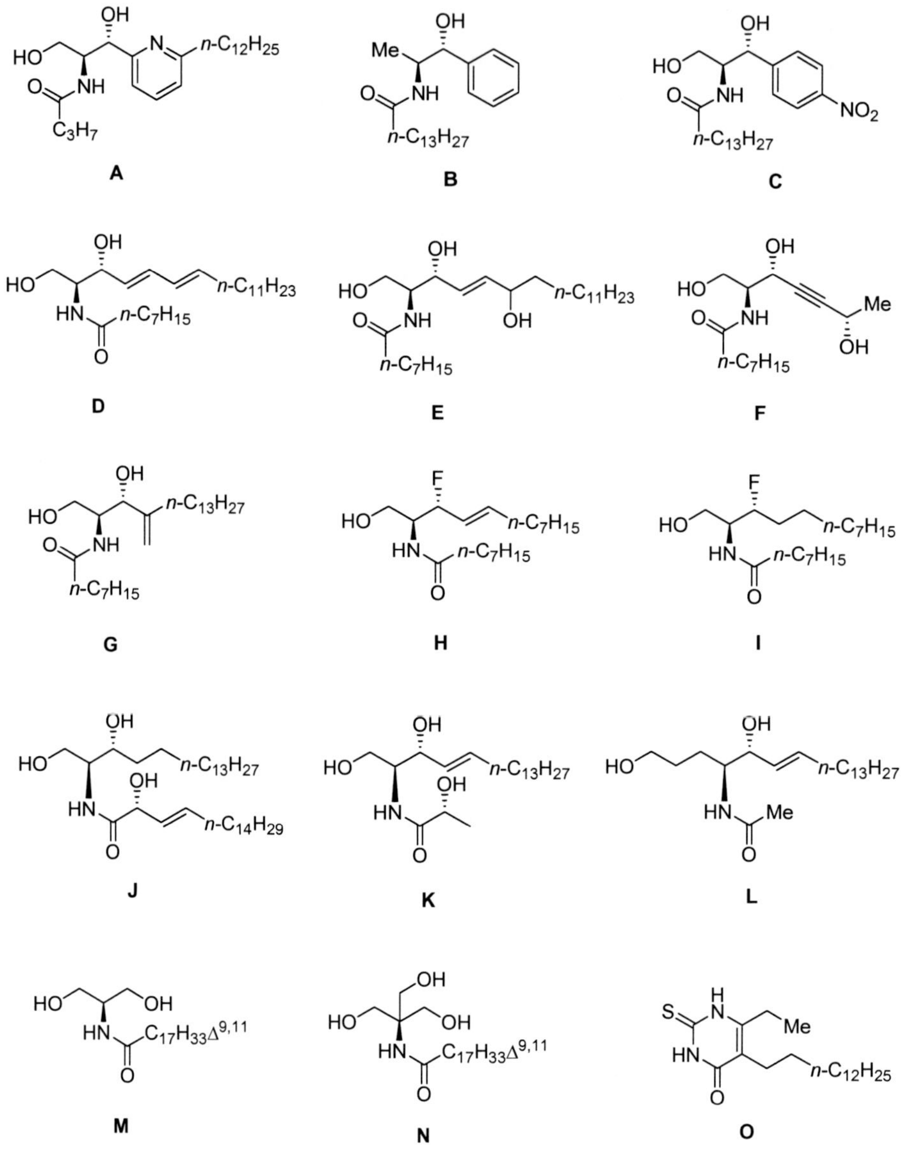

Chart. The known synthetic ceramide analogs (A-O).

Interesting analog was designed and synthesized by extending double bond in the natural substrate Δ^4 to two double-bonds at $\Delta^{4,6}$ (D) (Chun et al., 2002), putting an additional hydroxyl at C6 (E) (Chun et al., 2003), introducing a triple bond instead of the double bond Δ^4 (F) (He et al., 2000), and introducing an *exo*-methylene at C4 (G) (Lu et al, 2005).

Some of them were tested as potential chemotherapeutic agents in breast cancer (Struckhoff et al., 2004). Among them 4,6-diene-Cer was more potent than natural C2-Cer in cell death in MDA-MB-231 and NCI/ADR-RES breast cancer cell lines with elevation of ceramide levels in MCF-7 cells. Fluoride analogs (H and I) for the substituent of OH at C-3 of ceramide and dihydroceramide showed interesting biological activities including stimulation of axonal growth of hippocampal neurons by H and induction of apoptosis in Molt-4 and K-422 leukemia cells by I (O'Hagan and Rzepa, 1997; De Jonghe et al., 1999;

Van Overmeire et al., 2000). Naturally occurring symbioramide (J) bearing the *trans* allylic alcohol functional group in the fatty acid side chain instead of sphingoid backbone was also served as a good lead for the modification of the ceramide analogs (Azuma et al., 2003a). Potent antileukemic activity against HL-60 cells was observed by *N*-lactylsphingosine (K) that consisted by the natural sphingosine backbone and *N*-hydroxyalkyl group as symbioramide does (Azuma et al., 2000 and 2003b). Analogs were further flooded with mimics derived from *N*-acyl diethanolamine (M) and *N*-acyl 2-amino-2-hydroxymethyl-1,3-propandiol (N) (Bieberich et al., 2002). The compound (L) having different methylene space between the primary hydroxymethyl group and aminomethylene of sphingosine backbone was also prepared and tested against human leukemia (HL-60) cells *in vitro* (Shikata et al., 2002). Most of them were less active than C2-Cer. Furthermore, most of acyclic ceramide analogs bearing primary alcohol in sphingoid backbone perturbed or al least have a chance to perturb the metabolic pathways by its possible involvement in various biological reactions.

Constrained ceramide analogs free from primary alcohol may overcome this drawback without metabolic perturbation. A good example was shown by uracil ring compound (O) in which the polar portion of ceramide was replaced (Macchia, et al., 2001; Macchia, et al., 2003). Additional long and short alkyl chains were introduced to mimic the sphingoid backbone of ceramide at C-5 and C-6 of the uracil ring positions respectively. This compound has very potent growth-inhibitory effect of CCRF-CEM human leukemia cells to trigger apoptosis by the occurrence of DNA and nuclear fragmentation (Macchia, et al., 2001). Potent antitumor activity was also observed by the treatment of female CD-1 nu/nu athymic mice bearing a WiDr human colon xenograft with lower systematic toxicity. However, these uracil analogs do not bear an allylic alcohol system and do have aromaticity in the main ring part, which are quite different from the real conformational alignment of acyclic C2-Cer.

Recently we reported a short communication for the synthesis and biological activities of new constrained ceramide analogs which are closer to the real conformational and electronic properties of ceramide on the basis of oxazolidin-2-one ring (5) (Ha et al., 2006).

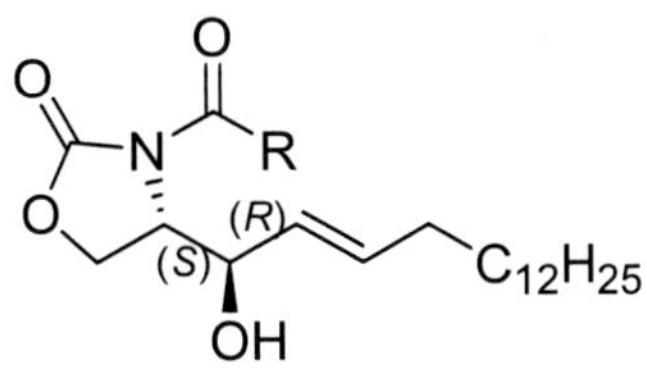

Constrained analogs (**5**)

Analogs with oxazolidin-2-one ring were designed to combine the primary alcohol at C1 and the amide nitrogen at C2 of ceramide together by additional carbonyl group. This conformation was deduced by early observation that there are strong intramolecular hydrogen-bonding among alcoholic hydrogens and the amide nitrogen (Rerek, et al., 2001). An analog, *N*-hexadecanoyl compound 5c, mimicked naturally occurring acyclic C16-Cer whose conformational similarity was evaluated by the comparison with the minimum energy conformers calculated by density functional model. The overlay structure of those two compounds, C16-Cer and 5c, shows similarity in Figure 1.

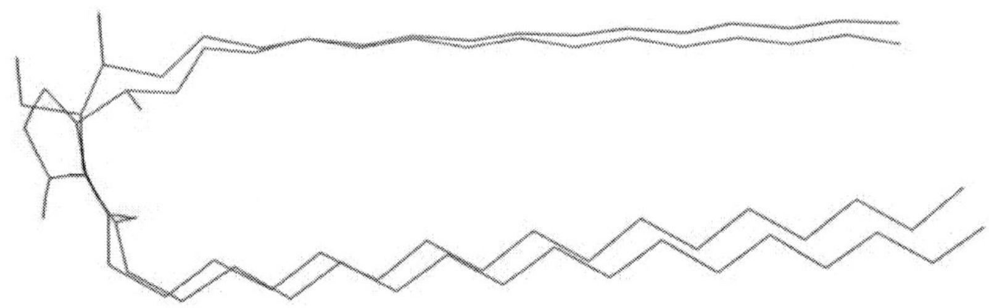

Figure 1. Overlay of naturally occurring C16-ceramide 1 and its oxazolidin-2-one analog 5c. Calculation was carried out with the commercial product Spartan® 04. The overlay structures were deduced by locking the bonds associated with nitrogen for comparison.

This analog retains the same conformation with the allylic alcohol system as for the natural substrate that is one of the essential structural elements to keep the biological activity with the generation of reactive oxygen species (Jarvis et al., 1996; Reynolds et al., 2004). Furthermore, in this analog the free primary hydroxyl group of the natural substrate was transformed to the oxazolidin-2-one ring that may block metabolic perturbation occurring at the hydroxyl group of C-1 of the sphingoid backbone.

All of the designed compounds were synthesized from commercially available (2*S*)-aziridine-2-carboxylate (7) shown in the following scheme.

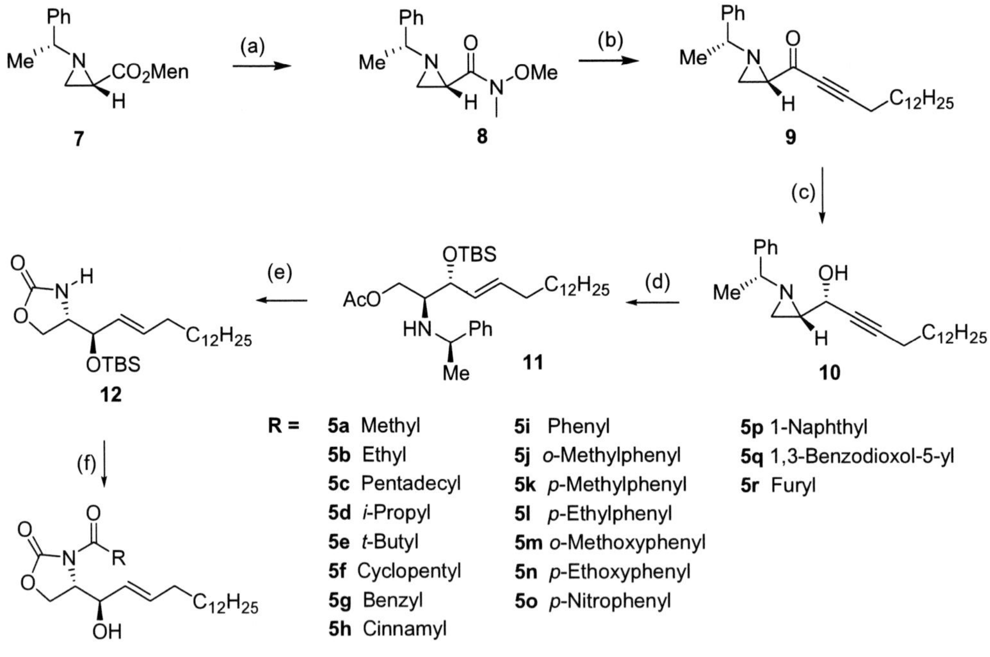

Scheme. (a) $NH(OMe)CH_3$, *i*-PrMgBr, THF, 0°C (92%); (b) $HCCHCH_2C_{12}H_{25}$, *n*-BuLi, -78°C then 0°C, 2h (99%); (c) $ZnCl_2$, $NaBH_4$, MeOH, -78°C, then -15°C, 1h (98%); (d) (i) LAH, 0°C, 60°C, 8h (86%); (ii) DMAP, *t*-$BuMe_2SiCl$, rt (93%); (iii) AcOH, CH_2Cl_2, rt, (73%); (e) (i) KOH, EtOH, rt, 1h (81%); (ii) CDI, DBU, rt, 24h (87%); (iii) Na, NH_3, -78°C (77%); (f) (i) RCOCl, NaHMDS, 0°C; (ii) Bu_4NF, THF, 0°C.

The aziridine ester 7 was reacted with *N,O*-dimethylhydroxylamine in the presence of *i*-PrMgBr at 0°C to form the Weinreb's amide 8 in 92% yield followed by addition of 1-pentadecynyl lithium to yield 9 in quantitative yield. Acyl aziridine 9 was selectively reduced to *threo*-amino alcohol 10 to have the right stereochemistry of ceramide with $NaBH_4$ and $ZnCl_2$ *via* the chelation-controlled transition state in 98% yield (Yun, 2003). Then alkyne 10 was reduced by LAH to yield *N*-[(*R*)-α-methylbenzyl]aziridin-2(*S*)-yl-hydroxy-1-pentadecene in 86% yield whose hydroxy group was protected by TBDMS in 93% yield. The aziridine ring was regioselectively opened by the reaction with acetic acid in CH_2Cl_2 to give acyclic compound 11 in 73%. The *O*-acetyl group in 11 was hydrolyzed by KOH in ethanol to yield the free alcohol at which stage oxazoldin-2-one ring was introduced by the reaction with 1,1'-carbodiimidazole in 70% yield for two steps. The α-methylbenzyl group attached on the nitrogen was removed by Na in liq. NH_3 at -78°C to give the backbone ring 12 in 77% yield. The parent ring (12) was readily acylated on the nitrogen using acid chloride in the presence of a suitable base. Sodium hexamethyldisilazane was the most effective base being applicable to various acyl chlorides. The target molecules (5a - 5r) were prepared from the parent ring compound 12 by the same reaction sequence with the corresponding acid chlorides followed by desilylations in over 78-55% yield in all cases.

All of these compounds were evaluated as potential antileukemic agents against human leukemia HL-60 cells with C2-Cer and C2-DHCer as positive and negative references respectively by MTT assay whose results were shown in Figure 2 (van de Loosdrecht et al., 1994). Human leukemia HL-60 cells were treated with 20 μM of each constrained ceramide analogs with references for 16 hours. All tested compounds (5a-5r) showed activities with drastic difference by the *N*-acyl substituents of the oxazolin-2-one ring. *N*-Acetyl compound 5a mimicking C2-Cer has much lower activity than C2-Cer itself while a great deal of improvement was observed in the compound 5b with additional one carbon as ethyl group in R of the compound 5. Compound 5c with a long alkyl chain did not show any activity due to low cell permeability to produce its antileukemic activity that was also observed in natural C16-Cer (Selzner et al., 2001; Fillet, et al., 2003). The longer carbon chain than ethyl (5b) such as *i*-propyl (5c) or butyl (5d) did not improve the activity much. The cyclorpropanyl ring at R in the compound 5f improved the activity drastically which showed the best activity among all tested compounds with alkyl substituents. Compounds with benzyl (5g) and cinnamoyl (5h) were not active. Some of the compounds possessing benzoyl (5i – 5o) rather than alkanoyl group (5a – 5h) showed good activities. Their activities were dependent to the substituents of phenyl ring. Especially the compound (5o) with *p*-nitrophenyl for R showed very potent activity. Aryl substituents other than phenyl for R group in the compounds 5 did not improve the activity among the tested compounds including naphthyl (5p), 1,3-benzodoxol-5-yl (5q) and furyl (5r) substituents. The origin of the activity difference among all tested compounds is speculated not from the simply physical properties such as lipophilicity or polarity but from the unique structural features of each compound.

We decided to find out whether the cytotoxicity induced by these analogs or the reduction of HL60 cells was caused by the apoptosis. The characteristics of cell death were verified by the analysis of DNA-fragmentation induced by these molecules. DNA-laddering of fragments was observed after 15h of incubation of HL-60 cells with 20 μM of C2-Cer, C2-

DHCer and oxazolidin-2-ones (5a, 5b, 5c, 5d, 5f, 5i and 5o) shown in Figure 3. The picture showed the clear co-relationship between anticancer activity and DNA-fragmentation.

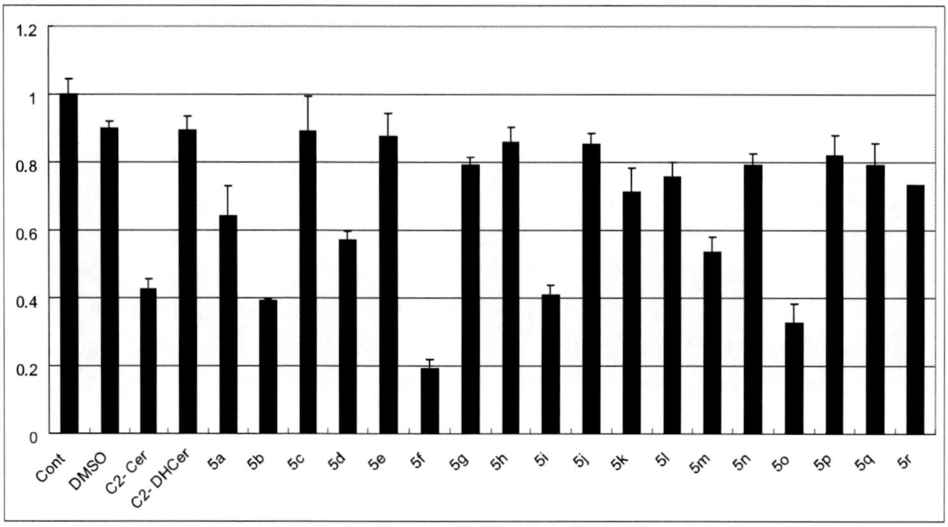

Figure 2. Relative cell survival of HL-60 cells after treatment with 20 μM of C2-Cer, C2-DHCer and oxazolidin-2-ones (5a - 5r) for 16 h.

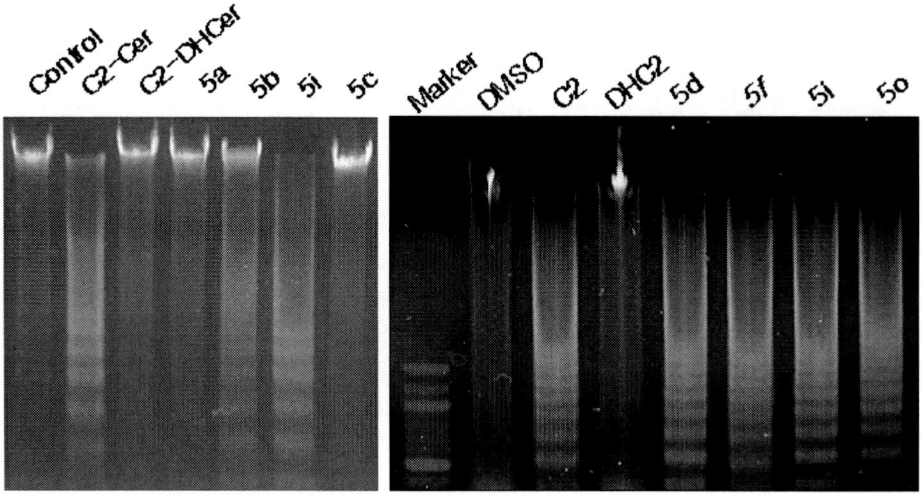

Figure 3. DNA fragmentation of HL-60 cells by the treatment with 20 μM of C2-Cer, C2-DHCer and some representative oxazolidin-2-ones (5a, 5b, 5c, 5d, 5f, 5i, 5o).

The active compounds such as 5b (R = ethyl) and 5i (R = phenyl) showed similar or a little more amount of DNA fragmentation as white bands than for C2-Cer while inactive 5a (R = methyl) and 5c (R = pentadecyl) had similar band to C2-DHCer's on the picture. All active compounds 5d (R = *i*-propyl), 5f (R = cyclopentyl), and 5o (*p*-nitrophenyl) also showed DNA-fragmented bands. This demonstrates cell death was caused by the apoptotic process. A good correlation between cell death of HL-60 cells and DNA fragmentation is a solid evidence that the cytotoxicity of the constrained ceramide analogs is originated from apoptosis.

The cellular concentration of ceramide, measured three times by the reported method (Dagan et al., 2003) after the treatment of the active analogs, showed minor changes by less than 15%, which means the oxazolidin-2-one ceramide analogs did not inhibit the ceramide metabolic enzymes to accumulate the cellular concentration of ceramide. Instead these molecules induced apoptosis through caspase cascade with induction of nuclear factor κB (NF- κB) DNA binding and caspase-3 activation as exogeneous C-2 and C-6-ceramides do (Fillet et al., 2003). The more detailed studies on the pro-apoptotic process and the structure and activity relationship with diverse oxazolidin-2-ones as constrained ceramide analogs are needed.

Acknowledgement

We gratefully acknowledge the financial support of the Korea Science and Engineering Foundation through the Center for Bioactive Molecular Hybrides.

References

Azuma, H.; Takao, R.; Niiro, H.; Shikata, K.; Tamagaki, S.; Tachibana, T.; Ogino, K. (2003a) Total syntheses of symbioramide derivatives from *L*-serine and their antileukemic activities. *J. Org. Chem. 68,* 2790-2797.

Azuma, H.; Takao, R.; Shikata, K.; Niiro, H.; Tachibana, Ogino, K. (2003b) Synthesis and potent antileukemic activities of *N*-lactylsphingosine and *N*-lactyldihydrosphingosine. *J. Med. Chem. 46*, 3445-3447.

Azuma, H.; Tamagaki, S.; Ogino, K. (2000) Stereospecific total syntheses of sphingosine and its analogues from *L*-Serine. *J. Org. Chem. 65*, 3538-3541.

Bieberich, E.; Hu, B.; Silva, J.; MacKinnon, S.; Yu, R. K.; Fillmore, H.; Broaddus, W. C.; Otterbrite, R. M. (2002) Synthesis and characterization of novel ceramide analogs for induction of apoptosis in human cancer cells. *Caner Lett. 181*, 55-64.

Bielawska, A. ; Greenberg, M. S. ; Perry, D.; Jayadev, S.; Shayman, J. A.; McKay, C.; Hannun, Y. A. (1996) (1S,2R)-*D*-Erythro-2-(*N*-myristoylamino)-1-phenyl-1-propanol as an inhibitor of ceramidase, *J. Biol. Chem. 271*, 12646-12654.

Bielawska, A. ; Linardic, C. M.; Hannun, Y. A. (1992) Ceramide-mediated biology. Determination of structural and stereospecific requirements through the use of *N*-acyl-phenylaminoalcohol analogs, *J. Biol. Chem. 267*, 18493-18497.

Brodeser, S.; Sawatzki, P.; Kolter, T. (2003) Bioorganic chemistry of ceramide. *Eur. J. Org. Chem. 61*, 2021-2034.

Burow, M. E.; Weldon, C. B.; Tang, Y.; Navar, G. L.; Krajewski, S.; Reed, J. C.; Hammond, T. G.; Clejan, S.; Beckman, B. S. (1998) Differencesin susceptibility to tumor necrosis factor alpha-induced apoptosis among MCF-7 breast cancer cell variants, *Cancer Res. 58*, 4940-4946.

Chang, Y. T.; Choi, J.; Ding, S.; Prieschl, E. E.; Baumruker, T.; Lee, J. A.; Chung, S. K.; Schultz, P. G. (2002) The synthesis and biological characterization of a ceramide library. *J. Am. Chem. Soc. 124,* 1856-1857.

Chun, J.; Byun, H.-S.; Bittman, R. (2003) First asymmetric synthesis of 6-hydroxy-4-sphingenine-containing ceramides. Use of chiral propargylic alcohols to prepare a lipid found in human skin. *J. Org. Chem. 68,* 348-354.

Chun, J.; He, L.; Byun, H.-S.; Bittman, R. (2000) Synthesis of ceramide analogues having the C(4)-C(5) bond of the long-chain base as part of an aromatic or heteroaromatic system. *J. Org. Chem. 65*, 7634-7640.

Chun, J.; Li, G.; Byun, H.-S.; Bittman, R. (2002) Synthesis of new trans double-bond sphingolipid analogues: 4,6 and 6 ceramides. *J. Org. Chem. 67,* 2600-2605.

Cifone, M. G.; Maria, R. D.; Roncaioli, P.; Rippo, M. R.; Azuma, M.; Lanier, L. L.; Santoni, A.; Testi, R. (1994) Apoptotic signaling through CD95 (Fas/Apo-1) activates an acidic sphingomyelinase. *J. Exp. Med. 180*, 1547-1552.

Cuadros, R. ; Montejo de Garcini, E.; Wandosell, F.; Faircloth, G.; Fernández-Sousa, J. M. ; Avila, J. (2000) The marine compound spisulosine, an inhibitor of cell proliferation, promotes the disassembly of actin stress fibers. *Cancer Lett. 152*, 23-29.

Dagan, A.; Wang, C.; Fibach, E.; Gatt, S. (2003) Synthetic, non-natural sphingolipid analogs inhibit the biosynthesis of cellular sphingolipids, elevate ceramide and induce apoptotic cell death. *Biochim. Biophys. Acta, 1633*, 161-169.

De Jonghe, S.; Van Overmeire, I.; Gunst, J.; De Bruyn, A.; Hendrix, C.; Van Calenbergh, S.; Busson, R.; De Keukeleire, D.; Philippe, J.; Herdewijn, P. (1999) Synthesis and apoptogenic activity of fluorinated ceramide and dihydroceramide analogues. *Bioorg. Med. Chem. Lett. 9,* 3159-3164.

Dickson, R. C. (1998) Sphingolipid functions in *saccharomyces cerevisae*: comparison to mammals. *Ann. Rev. Biochem. 67,* 27-48.

Fillet, M.; Bentires-Alj, M.; Deregowski, V.; Greimers, R.; Gielen, J.; Piette, J.; Bours, V.; Merville, M.-P. (2003) Mechanisms involved in exogenous C2- and C6-ceramide-induced cancer cell toxicity. *Biochem. Pharmacol. 65,* 1633-1642.

Ha, H.-J.; Hong, M. C.; Ko, S. W.; Kim, Y. W.; Lee, W. K.; Park., J. (2006) Synthesis of constrained ceramide analogs and their potent antileukemic activities. *Bioorg. Med. Chem. Lett. 16*, 1880-1883.

Haimovitz-Friedman, A.; Kan, C. C.; Ehleiter, D.; Persaud, R. S.; McLoughlin, M.; Fuks, Z.; Kolesnick, R. N. (1994) Ionizing radiation acts on cellular membranes to generate ceramide and initiate apoptosis. *J. Exp. Med. 180*, 525-535.

Hannun, Y. A. (1994) The sphingomyelin cycle and the second messenger function of ceramide. *J. Biol. Chem. 269*, 3125-3128.

Hannun, Y. A. (1996) Functions of ceramide in coordinating cellular responses to stress *Science 274*, 1855-1859.

Hannun, Y. A., Luberto, C, Argraves, K. M. (2001) Enzymes of sphingolipid metabolism: from modular to integrative signaling. *Biochem. 40*, 4893-4903.

He, L.; Byun, H.-S.; Bittman, R. (2000) Stereoselective preparation of ceramide and its skeleton backbone modified analogues via cyclic thionocarbonate intermediates derived by catalytic asymmetric dihydroxylation of α,β-unsaturated ester precursors. *J. Org. Chem. 65*, 7627-7633.

Jarvis, W. D.; Fornari, F. A., Jr.; Traylor, R. S.; Martin, H. A.; Kramer, L. B.; Erukulla, R. K.; Bittman, R.; Grant, S. (1996) Induction of apoptosis and potentiation of ceramide-mediated cytotoxicity by sphingoid bases in human myeloid leukemia cells *J. Biol. Chem. 271*, 8275-8284.

Karasavvas, N.; Erukulla, R. K.; Bittman, R.; Lockshin, R. (1996) Octanoyl sphingosine stereoisomers and *n*-octyl-sphingosine induce apoptosis with variable potency in U936 cells. The ceramide carbonyl group is not required for apoptosis. *Eur. J. Biochem. 236*, 729-737.

Kim, J.; Kim, Y.-W.; Inagakic, Y.; Hwanga, Y.-A.; Mitsutakec, S.; Ryue, Y.-W.; Lee, W.-K., Ha, H.-J.; Park, C.-S.; Igarashi, Y. (2005) Synthesis and evaluation of sphingoid analogs as inhibitors of sphingosine kinases. *Bioorg. Med. Chem. 13*, 3475-3485.

Kim, K.; Kim, Y.-L.; Sacket, S. J.; Kim, H.-L.; Han, M.; Park, D. S.; Lee, B. K.; Lee, W. K.; Ha, H.-J.; Im, D.-S. (2007) Sphingosine 1-phospahte (S1P) induced sharp change in rat C6 glioma cells through $S1P_2$ receptor: development of an agonist for S1P receptor. *J. Pharm. Pharmacol. 59*, 1035-1041.

Kim, M.; Linardic, C.; Obeid, L.; Hannun, Y. (1991) Identification of sphingomyelin turnover as an effector mechanism for the action of tumor necrosis factor alpha and gamma-interferon. Specific role in cell differentiation. *J. Biol. Chem. 266*, 484-489.

Kinloch, R. A.; Treherne, J. M.; Furness, L. M.; Hajimohamadreza, I. (1999) The pharmacology of apoptosis, *Trends Pharmacol. Sci. 20*, 35-42.

Kolter, T.; Sandhoff, K. (1996) Inhibitors of glycosphingolipid biosythesis. *Chem. Soc. Rev. 25*, 371-381.

Kolter, T.; Sandhoff, K. (1999) Sphingolipids - Their metabolic pathways and the pathobiochemistry of neurodegenerative diseases. *Angew. Chem. Int. Ed., 38,* 1532-1568.

Liao, J.; Tao, J.; Lin, G.; Liu, D. (2005) Chemistry and biology of sphingolipids. *Tetrahedron, 61*, 4715-4733.

Liu, H.; Toman, R. E.; Goparaju, S. K.; Maceyka, M.; Nava, V. E.; Sankala, H.; Payne, S. G.; Bektas, M.; Ishii, I.; Chun, J.; Milstien, S.; Spiegel, S. (2003) Sphingosine kinase type 2 is a putative BH3-only protein that induces apoptosis. *J. Biol. Chem. 278*, 40330–40336.

Lu, X.; Arthur, G.; Bittman, R. (2005) Synthesis of a novel ceramide analogue via Tebbe methylenation and evaluation of its antiproliferative activity. *Org. Lett. 7*, 1645-1648.

Macchia, M; Barontini, S.; Bertini, S.; Di Bussolo, V.; Fogli, S.; Giovannetti, E.; Grossi, E.; Minutolo, F.; Danesi, R. (2001) Design, synthesis, and characterization of the antitumor activity of novel ceramide analogues. *J. Med. Chem. 44*, 3994-4000.

Machida, M.; Bertini, S, Fogli, S.; Giovannetti, E.; Minutolo, F.; Rapposelli, S.; Danesi, R. (2003) Ceramide analogues in apoptosis: a new strategy for anticancer drug development. *Il Farmaco 58*, 205-211.

O'Hagan, D.; Rzepa, H. S. (1997) Some influences of fluorine in bioorganic chemistry. *Chem. Commun.* 645-652.

Pettus, B. J., Chalfant, C. E.; Hannun, Y. A. (2002) Ceramide in apoptosis: an overview and current perspectives. *Biochim. Biophys. Acta. 1585*, 114-125.

Radin, N. S. (2003) Designing anticancer drugs via the achilles heel: ceramide, allylic ketones, and mitochondria. *Bioorg. Med. Chem. 11,* 2123-2142.

Rajewski, R. A.; Kosednar, D. G.; Matches, T. A.; Wong, O. S.; Burchett,. K.; Thakker, K. (1995) Stereospecific analysis of a novel protein kinase C inhibitor. *J. Pharm. Biomed. Anal. 13*, 247-253.

Rerek, M. E.; Chen, H.-c.; Markovic, B.; Wyck, D. V.; Garidel, P.; Mendelsohn, R.; Moore, D. J. (2001) Phytosphingosine and sphingosine ceramide headgroup hydrogen bonding: structural insights through thermotropic hydrogen/deuterium exchange. *J. Phys. Chem. B. 105*, 9355-9362.

Reynolds, C. P.; Maurer, B. J.; Kolesnick, R. N. (2004) Ceramide synthesis and metabolism as a target for cancer therapy. *Cancer Lett. 206,* 169-180.

Sachs, C. W.; Safa, A. R.; Harrison, S. D.; Fine, R. L. (1995) Partial inhibition of multidrug resistance by safingol is independent of modulation of P-glycoprotein substrate activities and correlated with inhibition of protein kinase C. *J. Bol. Chem. 270*, 26639-26648.

Sawai, H.; Hannun, Y. A. (1999) Ceramide and sphingomyelinases in the regulation of stress responses. *Chem. Phys. Lipids 102*, 141-147.

Selzner, M.; Bielawska, A.; Morse, M. A.; Rudiger, H. A.; Sindram, D.; Hannun, Y. A.; Clavien, P. A. (2001) Induction of apoptotic cell death and prevention of tumor growth by ceramide analogues in metastatic human colon cancer. *Cancer Res. 61*, 1233-1240.

Shikata, K.; Azuma, H.; Tachibana, T.; Ogino, K. (2002) Synthesis of novel and non-natural ceramide analogues derived from glutamic acid *Tetrahedron, 58*, 5803-5809.

Struckhoff, A. P.; Bittman, R.; Burow, M. E.; Clejan, S.; Elliott, S.; Hammond, T.; Tang, Y.; Beckman, B. S. (2004) Novel ceramide analogs as potential chemotherapeutic agents in breast cancer. *J. Pharmacol. Exp. Ther. 309,* 523-532.

Van de Loosdrecht, A. A.; Beelen, R. H. J.; Ossenkoppele, G. J.; Broekhoven, M. G.; Langenhuijsen, M. M. A. C. (1994) A tetrazolium-based colorimetric MTT assay to quantitate human monocyte mediated cytotoxicity against leukemic cells from cell lines and patients with acute myeloid leukemia. *J. Immunol. Methods 174*, 311-320.

Van Overmeire, I.; Boldin, S. A.; Venkataraman, K.; Zisling, R.; De Jonghe, S.; Van Calenbergh, S.; De Keukeleire, D.; Futerman, A. H.; Herdewijn, P. (2000) Synthesis and biological evaluation of ceramide analogues with substituted aromatic rings or an allylic fluoride in the sphingoid moiety. J. *Med. Chem. 43*, 4189-4199.

Wang, C.Y. ; Cusack, Jr., J. C.; Liu, R.; Baldwin, Jr., A. S. (1999) Control of inducible chemoresistance: enhanced anti-tumor therapy through increased apoptosis by inhibition of NF-kappaB, *Nat. Med. 5*, 412-417.

Yun, J. M.; Sin, T. B.; Hahm, H. S.; Lee, W. K.; Ha, H.-J. (2003) Efficient synthesis of enantiomerically pure 2-acylaziridines: facile syntheses of *N*-Boc-safingol, *N*-Boc-D-erythro-sphinganine, and *N*-Boc-spisulosine from a common intermediate. *J. Org. Chem. 68*, 7675-7680.

In: Glycolipids: New Research
Editor: Daiki Sasaki, pp. 157-169

ISBN: 978-1-60456-216-3

Chapter VIII

Pivotal Roles of Cholesterol and Gangliosides in Aggregation of Amyloid β-Protein

Katsuhiko Yanagisawa
Department of Alzheimer's Disease Research,
National Institute for Longevity Sciences, National Center
for Geriatrics and Gerontology, 36-3 Gengo,
Morioka, Obu 474-8522, Japan

Abstract

Increasing lines of evidence suggest that cholesterol is a potent risk factor for the development of Alzheimer's disease (AD). An increase in the cholesterol level of neuronal membranes likely enhances both the generation and aggregation of the amyloid β-protein (Aβ). Regarding the enhancement of Aß aggregation, previous studies by our group and other groups suggest that cholesterol has both direct and indirect effects on such aggregation. We previously identified a novel Aβ species in the brain exhibiting early pathological changes of AD. On the basis of unique molecular characteristics, we hypothesized that GM1 ganglioside (GM1) plays a critical role in the pathological processes in AD. The mechanism underlying the binding of Aβ to GM1 ganglioside remains to be determined; however, our recent studies suggest that an increase in the cholesterol level in the membrane facilitates such binding. The distribution of cholesterol in neuronal membranes is likely to be altered by aging and the expression of apolipoprotein E4, both of which are strong risk factors for AD development. The putative roles of cholesterol and gangliosides in AD development are discussed.

1. Introduction

Amyloid β–protein (Aβ) deposition in the brain is one of the fundamental processes in the development of Alzheimer's disease (AD); however, the molecular mechanism underlying the conversion of monomeric, nontoxic Aβ to its aggregated toxic forms remains to be determined. Aβ, a physiological derivative of the amyloid precursor protein (APP), is prone to aggregate by itself at its extremely high concentrations. However, Aβ aggregation *in vitro* can be significantly suppressed by subjecting an Aβ solution to "de-seeding" by ultracentrifugation to remove undissolved peptides (Naiki and Gejyo, 1999). The expression of genes responsible for familial AD is likely to accelerate Aβ deposition through enhanced Aβ generation (Selkoe, 1997). However, in the case of sporadic AD, a major form of the disease, no evidence has been reported to date to indicate that Aβ generation is enhanced. Furthermore, accumulating evidence suggests that *APP* gene mutations within Aβ sequence induce Aβ aggregation and deposition in brain without enhanced Aβ generation (Nilsberth et al., 2001). Thus, it is reasonable to assume that Aβ deposition in the brain is induced not only by enhanced Aβ generation but also by as yet unclarified environmental factors. We previously hypothesized that Aβ starts to aggregate and deposit in the brain through its binding to GM1 ganglioside (GM1) (Yanagisawa, et al., 1995: Yanagisawa et al, 1998). To date, evidence that supports our hypothesis is growing from in vitro and in vivo studies (McLaurin and Chakrabartty, 1996; Choo-Smith and Surewicz, 1997; Choo-Smith et al., 1997; Matsuzaki and Horikiri, 1999, Kakio et al., 2001, 2002; Hayashi et al., 2004). We also previously found that binding of Aβ to GM1 is facilitated by an increased cholesterol level in neuronal membranes (Kakio et al., 2001).

2. Cholesterol and Gangliosides in Brain

Cholesterol is one of the major lipids present in cellular membranes and is essential for many cellular functions. This is also the case for cholesterol in the brain; however, little is known about cholesterol metabolism in the brain. The brain is highly rich in cholesterol. It accounts for only 2% of the whole body mass but contains almost one-quarter of the unesterified cholesterol in the whole body (Dietschy and Turley, 2001). A neuron is characterized by its specific synaptic plasma membrane. Notably, cholesterol accounts for more than 40 mol% of the total membrane lipids in synaptic plasma membranes (Schroeder et al., 1991; Wood et al., 1989). Thus, it is important to clarify how cholesterol in neuronal membranes, particularly in synaptic plasma membranes, is regulated.

Although a recent study has indicated the possibility of cholesterol transport across the blood-brain barrier (Refolo et al., 2000), it is likely that cholesterol turnover is restricted only inside the brain. at a very low rate. Only 0.02% of the cholesterol pool in the brain undergoes a turnover each day (Dietschy and Turley, 2001). This is less than 1% as fast as the cholesterol turnover in the whole body (Dietschy and Turley, 2001). These lines of evidence indicate that the function and metabolism of cholesterol in the brain are very unique. Most cholesterol in the brain is likely to come from *in situ* synthesis; however, the rate of *in situ* cholesterol synthesis in many organs, including the nervous system, depends on age (Cenedella and Shi, 1994; Stahlberg et al., 1991; Goodrum, 1990; Popplewell and Azhar, 1987). Moreover, it was previously reported that cholesterol is the only major lipid in the brain that cannot be synthesized at the ends of neurites (Vance et al., 1994), indicating that the cholesterol level at the ends of neurites depends on two inputs: axonal flow from the soma and the uptake of extracellular cholesterol. Thus, it is likely that neuronal functions in the elderly depend more on exogenous cholesterol level through apoE receptors. Moreover, aging and apoE4 expression (Corder eta., l1993; Poirier et al., 1993; Saunders et al., 1993; Strittmatter et al., 1993) are strong risk factors for AD development. Thus, it is important to elucidate how aging- and apoE4-dependent alteration in cholesterol metabolism accelerates Aβ aggregation and deposition in the brain.

3. Cholesterol-Dependent Acceleration of Aß Aggregation and Deposition in Brain

3.1. Possibility of Direct Acceleration of Aβ Aggregation by Cholesterol

The interaction between Aβ and plasma membranes is important subject for elucidating various Aβ-induced pathological processes. The possibility of the direct binding of Aβ to lipids, such as cholesterol, phosphatidylcholine and fatty acids, was previously investigated (Avdulov et al., 1997). Interestingly, it was shown that the Aβ binding to lipids is dependent on the aggregation state of Aβ, that is, monomeric Aβ does not bind to lipids and only aggregated Aβ binds to lipids in a lipid-class-dependent manner. Alternatively, it was also suggested that Aβ-lipid interaction is modulated by cholesterol level in neuronal membranes using various techniques: that is, the level of Aβ binding to the surface of cultured PC 12 cells decreases with increasing cholesterol level in the cells and that cholesterol-depleted cells has a higher extent of Aβ-cell surface binding (Yip et al., 2001). It was also shown that cholesterol content in neuronal membranes negatively correlates with the membrane perturbation effect of Aβ (Kirsch et al., 2002). This evidence suggests that Aβ interaction with neuronal membrane is significantly modulated by cholesterol content in such membranes. However, regarding the potency of cholesterol in accelerating or inhibiting Aβ fibrillogenesis, the conclusions drawn in previous studies were controversial. In one of such studies, it was investigated how Aβ fibrillogenesis is modulated in the presence of cholesterol by transmission electron microscopy using aqueous suspensions of microcrystalline cholesterol and other lipids, and suggested that the marked acceleration of Aβ aggregation is induced in the presence of cholesterol (Harris, 2002). Furthermore, this study suggests that the potentiation of Aβ polymerization is likely to be induced by the hydrophobic interaction between the amino acid side chains of Aβ and the tetracyclic sterol nucleus. In contrast, another study suggested on the basis of the results of *in situ* atomic force microscopy and *ex situ* electron microscopy that bilayer cholesterol content in total brain lipid extracts inversely correlates with the extent of Aβ insertion into neuronal membranes and subsequent Aβ fibrillogenesis on the neuronal membrane surface (Yip et al., 2001). Regarding the modulation of Aβ insertion into membranes by cholesterol, it was previously shown that the extent of Aβ insertion into lipid bilayers rather increases as cholesterol content increases, and such insertion subsequently inhibits Aβ fibrillogenesis on the neuronal membrane surface (Ji et al., 2002). Thus, it is very difficult at this point to conclude whether an increase in the cholesterol content of neuronal membranes directly accelerates or inhibits Aβ aggregation.

3.2. Cholesterol-Dependent Generation of Seed for Aβ Aggregation

3.2.1. Identification of GM1-Ganglioside-Bound Aβ

To clarify the molecular mechanism underlying Aβ aggregation and deposition in the brain, we previously attempted to identify an Aβ species that initially accumulates in the brains of AD patients (Yanagisawa et al., 1995). In this study, we identified a novel Aβ species characterized by its tight binding to GM1 only in the brains that exhibit early pathological changes of AD. The GM1-ganglioside-bound Aβ (GAβ) showed unique molecular characteristics, including a markedly high aggregation potential and an altered immunoreactivity (Yanagisawa et al., 1995; Yanagisawa et al., 1997). On the basis of these molecular characteristics, we hypothesized that Aβ adopts an altered conformation through binding to GM1, and then acts as a seed for Aβ aggregation in the brain. Subsequent to our initial report on GAß, several investigators performed *in vitro* studies, and the results of their studies support our hypothesis.

3.2.2. Generation of Monoclonal Antibody Specific to Gaβ

To test our hypothesis, we previously attempted to generate monoclonal antibodies specific to the altered conformation of Aβ (Yanagisawa et al., 1997). We prepared a neuronal membrane fraction from the cerebral cortices of subjects with abundant diffuse plaques as described above and then purified GAβ from the neuronal membrane fraction containing GAβ were concentrated using a Microcon 3 concentrator (Amicon). To raise monoclonal antibody, we incubated the purified GAβ with cultured spleen cells obtained from mice for *in vitro* immunization. The incubated spleen cells were hybridized with myeloma cells in an electric fusion apparatus. Positive clones were obtained through the screening of the supernatants of hybridoma cell cultures by Western blotting and the immunostaining of the membrane fraction smear on glass slides. By these procedures, we first obtained an IgM monoclonal antibody (4396) (Yanagisawa et al., 1997). From the original IgM hybridoma cells, the IgG monoclonal antibody (4396C) was produced by the genetic class-switch technique (Hayashi et al., 2004). The specificity of 4396C binding to GAβ was confirmed by quantitative binding assay and dot blot analysis of liposomes carrying GAβ on their surface. In this experiment, 4396C did not react with the free forms of Aβ and GM1 (Hayashi et al., 2004). 4396C recognized the GM1-bound forms of two Aβ isoforms (Aβ1-40 and Aβ1-42). In contrast, 4G8, a monoclonal aqntibody specific to the midregion of Aβ, did not react with GAβ (Hayashi et al., 2004). From this result, we concluded that the midregion of Aβ, but not the amino or carboxyl terminus, is altered through its binding to GM1.

3.2.3. Confirmation of the GAβ Generation in Brain

Having established the specific binding of 4396C to GAβ, we attempted to confirm GAβ generation in the brain. We performed the immunohistochemistry of GAβ in sections of AD brains. However, we did not detect any 4396C immunoreactivity in the sections fixed in formaldehyde and pretreated with formic acid to enhance Aβ immunoreactivity (Hayashi et al., 2004). This result suggests that the conformation of GAβ is sensitive to routine immunohistochemical procedures. We then searched for an alternative procedure that allows us to detect GAβ by immunohistochemistry and found that fixation with Kryofix was useful to detect GAβ. With the Kryofix fixation and SDS-pretreatment of the sections, neurons in the AD brain sections were immunostained by 4396C (Hayashi et al., 2004). In contrast to the immunostaining by 4396C, 4G8 immunostained plaques but not neurons under the same conditions. This result again suggests that the midregion of Aβ is significantly altered through its binding to GM1. We examined many brain sections obtained at necropsy of patients with AD and those with Down syndrome (DS), and of individuals who died without any evidence of dementia. The most strong neuronal staining by 4396C was observed in the brains of DS patients. Neuronal staining by 4396C was absent or at levels comparable to those of AD brains in control brains. From these results, we concluded that GAβ can be detected by immunohistochemistry, and that neuronal immunostaining by 4396C can also be nonspecifically induced. Since the nonspecific immunostaining of 4396C was likely due to changes that occur during the agonal and/or postmortem state, we examined fresh brains of nonhuman primates, which naturally develop Aβ deposition after 25. We performed immunohistochemistry of the cortical sections obtained from seven animals aged 4, 4, 5, 17, 19, 30 and 36. In the sections obtained from the two oldest animals (30 and 36), many neurons were strongly immunostained by 4396C with a granular pattern (Hayashi et al., 2004). These neurons were also recognized using a monoclonal antibody specific to the amino terminus of Aβ and cholera toxin, which is a natural ligand for GM1 (Hayashi et al, 2004). Furthermore, GAβ was immunoprecipitated only from the oldest animals by 4396C (Hayashi et al., 2004). These results suggested that GAβ is generated in the brain.

3.2.4. Molecular and Biological Mechanisms Underlying GAβ Generation

It remains to be determined how GAβ is generated in the brain, specifically in the AD brain. Aβ is physiologically generated from the amyloid precursor protein (APP) and secreted into extracellular space. GM1 is also physiologically expressed throughout the central and peripheral nervous systems. However, GAβ can be detected only in brains that show Aβ deposition (Yanagisawa et al., 1995). Matsuzaki and his colleagues previously attempted to elucidate the molecular mechanism underlying Aβ binding to GM1. They performed physicochemical experiments using liposomes with various compositions of lipids and found that increases in not only GM1 content but also cholesterol content in these liposomes accelerates Aβ binding to GM1 (Kakio et al., 2001). Regarding the mechanism underlying the cholesterol-dependent acceleration of Aβ binding to GM1, they suggested that GM1 clustering, which is likely to be recognized by Aβ as receptor, is induced in a cholesterol-rich environment (Kakio et al., 2001). It remains to be clarified how the distribution and content of cholesterol are regulated in neuronal membranes; however, a previous study showed that cholesterol distribution throughout the lipid bilayers of the synaptic plasma membrane (SPM) changes with age, resulting in an approximately twofold increase in cholesterol level in the exofacial leaflet of SPM prepared from old mice compared with that prepared from young mice (Igbavboa et al., 1996). We also previously analyzed cholesterol distribution in the SPM of human apoE-knock-in mice and found that apoE4 knock-in mice showed an approximately twofold increase in cholesterol level in the exofacial leaflet compared with those in the SPM of apoE3 knock-in mice and wild-type mice (Hayashi et al., 2002). Aging and apoE4 expression are strong risk factors for AD development; thus, a change in lipid level in neuronal membranes may be critical for the induction of Aβ assembly and its deposition in the brain.

3.2.5. Implications of GAβ Generation in AD Development

The seeded polymerization theory was previously proposed for the general process of pathological assembly and the deposition of constituent proteins in the brain, including Aβ and the prion protein (Harper and Lansbury, 1997). However, it remains to be determined how a soluble, nontoxic monomeric protein is converted to its aggregated or insoluble, toxic forms. In the aggregation of the prion protein, it was previously suggested that a seed for the assembly of the prion protein is generated by the binding of a monomeric prion protein to another molecule, *protein X* (Telling et al., 1995). The results of our previous studies support the seeded polymerization theory and, to the best of our knowledge, the detection of GAβ is the first evidence of the endogenous generation of a seed for Aβ assembly.

We have recently extended our study on GM1-induced Aβ aggregation using the hereditary Aβ variants. We examined how the aggregation of the Arctic- and Dutch-type Aβs were accelerated in the presence of GM1. Interestingly, these mutations cause the replacement of an amino acid at the same position; however, the Arctic-type mutation induces Aβ deposition exclusively in the brain parenchyma, leading to development of early-onset AD whereas the Dutch-type mutation predominantly induces Aβ deposition in cerebral blood vessel walls, leading to cerebral amyloid angiopathy. Our previous studies suggested that GM1 plays a critical role in Aβ aggregation and deposition in the AD brain; thus, we hypothesized that the aggregation of the Arctic-type Aβ but not the Dutch-type Aβ was accelerated in the presence of GM1. Indeed, in the experiment using liposomes containing GM1, significant Aβ aggregation is only observed with the Arctic-type Aβ but not with the Dutch-type Aß (Yamamoto et al., 2004a). This result suggests, for the first time, that area-specific Aβ deposition is likely induced by local gangliosides in the brain.

4. Alteration in Contents and/or Distributions of Gangliosides and Cholesterol in AD Brain

Regarding the mechanism underlying GAβ generation in the brain, we hypothesized that the changes in the content and/or distribution of gangliosides and cholesterol are induced by risk factors for AD development, including aging and apoE4 expression. To examine this possibility, we first investigated whether the GM1 content changes with age and/or by apoE4 expression. We analyzed apoE3- or apoE4-knock-in mice at three different ages: 2 months, 1 year and 2 years old (Yamamoto et al., 2004b). In this study, we focused on detergent-resistant membrane microdomains (DRMs) since DRMs are rich in gangliosides and cholesterol. GM1 level in DRMs obtained from synaptosomes increased with age, and this increase was more marked in apoE4-knock-in mouse brains than the apoE3-knock-in mouse brains (Fig. 1). Interestingly, further characterization of the age-dependent increase in the GM1 level of DRMs suggested that GM1-accumulated DRMs are distinct from lipid rafts for the following reasons. First, there is no apparent age-dependent increase in the levels of BCθ, which specifically binds to cholesterol-rich microdomains such as lipid rafts (Simon and

Ikonen, 1997), and antibodies specific to flotillin and the prion protein, which are selectively localized in lipid rafts (Bickel et al., 1997; Vey et al., 1996) (Figs. 2A, 2B). It remains to be elucidated how GM1 is accumulated in such unique microdomains of the synaptic plasma membrane in an age-dependent manner and an apoE4 expression-dependent manner. However, we investigated whether an increase in GM1 level facilitates Aβ aggregation *in vitro*. In these experiments, the acceleration of Aβ aggregation was observed in the incubation with the synaptosomes obtained from aged apoE4-expressing mice (Figs. 3A, 3B). This study, for the first time, provided evidence that aging and apoE4 expression synergistically accelerate Aβ aggregation and deposition in the brain through an increase in GM1 content in neuronal membrane microdomains. However, it remains unclarified how GM1 is accumulated in such unique microdomains in neuronal membranes under these conditions. One of the possible explanations is that GM1 in such specific microdomains is stabilized by an unknown mechanism; for example, an alteration in the local cholesterol level, which is induced by risk factors for AD development. Although further studies are needed, our studies suggest that an increase in local GM1 content is a prerequisite for the pathological aggregation and deposition of Aβ in the brain. This possibility has recently been supported by a study of autopsy brains of patients with AD (Molander-Melin et al., 2005). In this study, it was found that DRMs isolated from the frontal cortices, which likely present early pathological changes of AD, contained significantly higher levels of GM1 and GM2 compared with those isolated from the brains of age-matched controls.

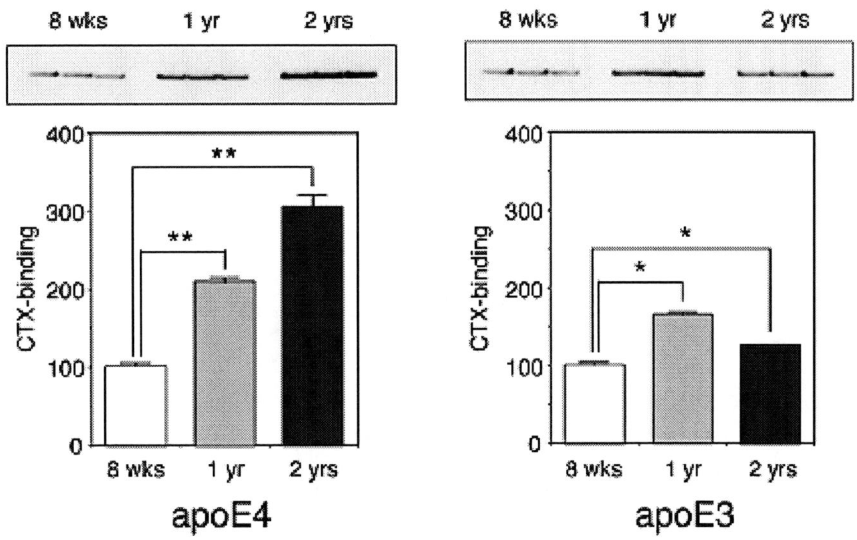

Figure 1. Binding of cholera toxin subunit B (CTX) to the blots of DRM fractions isolated from brain synaptosomes from three different age groups of apoE3- and apoE4-knock-in mice. Each lane contains 0.1 μg protein of DRM fraction isolated from synaptosomes prepared from combined five brains. The blots were reacted with HRP-conjugated CTX. The intensities of the bands relative to those for young (8 weeks old mice) are indicated. Each column represents the average value ± 1S.D. *: $p<0.0001$, **: $p<0.001$ (One-way ANOVA combined with Sheffe's test) (Reproduced from Yamamoto et al., FEBS Lett 569: 135-139, 2004).

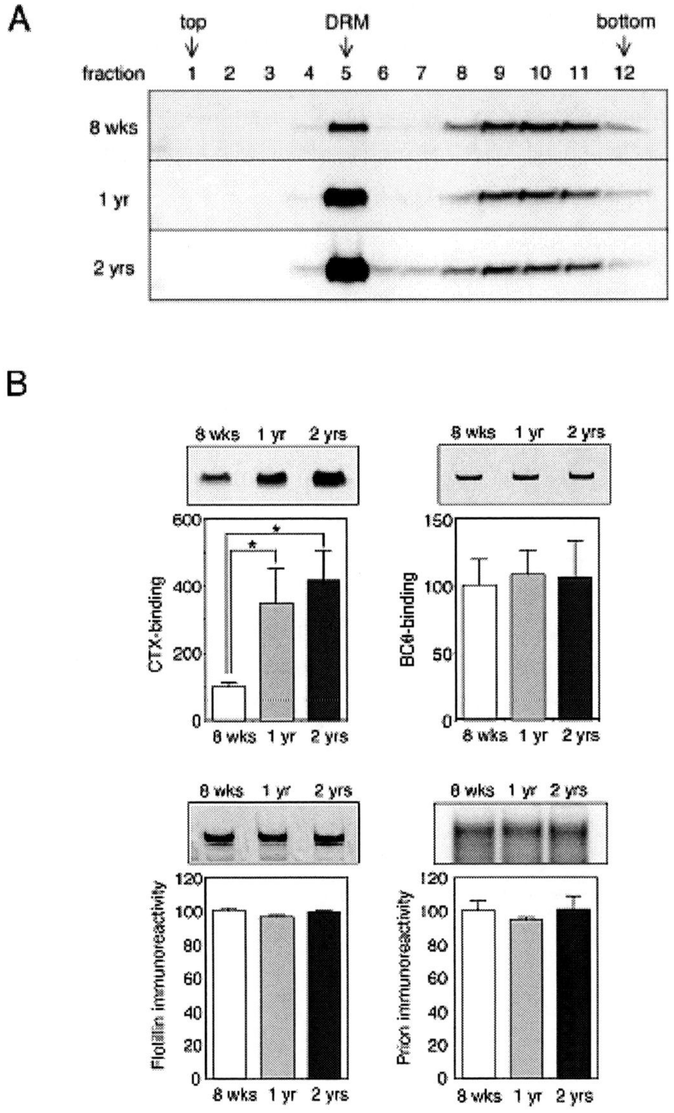

Figure 2. Binding of CTX-, BCθ- and anti-flotillin-antibody to the sucrose density gradient fractions, including DRMs, isolated from brain synaptosomes from three different age goups of apoE4-knock-in mice. (A) CTX binding to the blots containing all fractions of synaptosomes. (B) Binding of CTX and antibodies against flotillin and prion protein to the blots of DRM fractions. To examine the levels of BCθ bound to DRMs, synaptosomes were incubated with BCθ prior to sucrose density gradient fractionation. The relative intensities of the bands compared with those for 8 weeks old mice are indicated. Each column represents the average value ± 1S.D. *: $p<0.001$ (One-way ANOVA combined with Sheffe's test) (Reproduced from Yamamoto et al., FEBS Lett 569: 135-139, 2004).

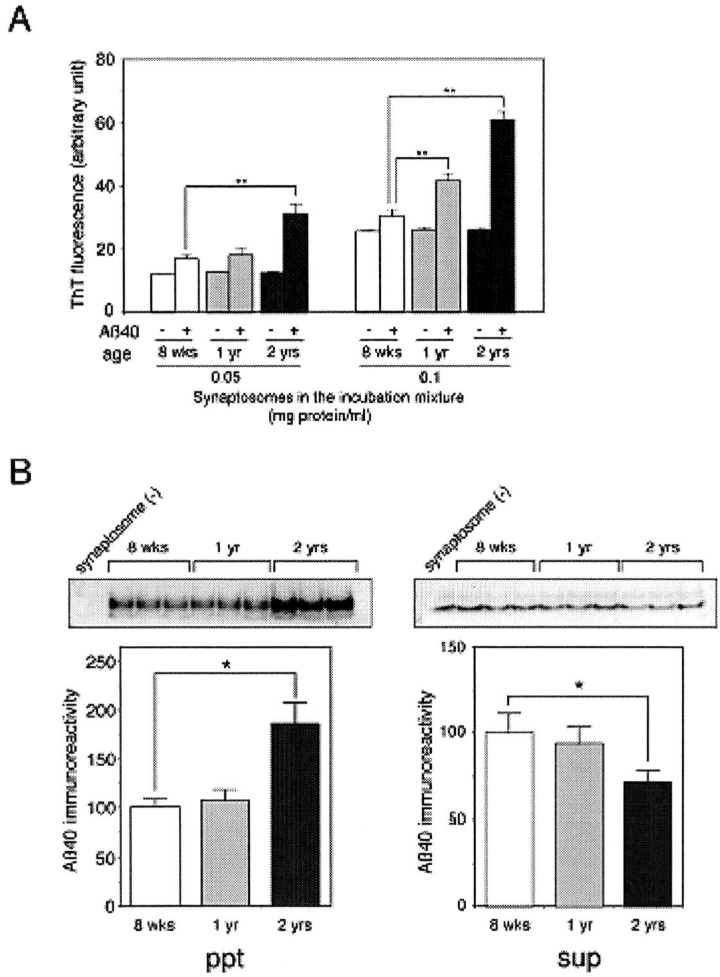

Figure 3. Accelerated Aβ aggregation in the presence of synaptosomes prepared from brains of the aged apoE4-knock-in mice. Aβ solutions were incubated at 50 μM and 37°C for 24 hr in the presence or absence of the synaptosomes isolated from three different age groups of mice (0.1 mg protein/ml). (A) ThT fluoresence intensities of the incubation mixtures. (B) Western blots of the precipitates (ppt) and supernatants (sup) of the incubation mixtures. The precipitated Aβ obtained from 100 μl of incubation mixture was subjected to electrophoresis following being solubilized in formic acid. The blots were reacted with 4G8, a monoclonal antibody specific to Aβ. The relative intensities of the bands compared with those in 8 weeks old mice are indicated. Each column represents the average value ± 1S.D. *: $p<0.05$ (One-way ANOVA combined with Sheffe's test) (Reproduced from Yamamoto et al., FEBS Lett 569: 135-139, 2004).

5. Conclusions and Perspectives

Considerable attention has been focused on the possible involvement of cholesterol in AD development (Ehehalt et al., 2003; Fassbender et al., 2001; Frears et al., 1999; Jick et al.,

2000; Kivipelto et al., 2001; Mizuno et at., 1999; Puglielli et al., 2003; Refolo et al., 2000; Simon et al., 1998; Wahrle et al., 2002; Wolozin et al., 2000). However, it remains to be determined how cholesterol accelerates the pathological processes of AD, namely, accelerated Aβ aggregation and deposition in the brain. To date, there are increasing lines of evidence suggesting that both the generation and aggregation of Aβ. is accelerated in a cholesterol-rich environment. Our previous studies suggest that cholesterol indirectly accelerates Aβ aggregation through induction of seed Aβ, that is, GAβ. Future studies are required to elucidate the molecular mechanisms underlying the generation of GAβ, particularlly in a cholesterol-rich environment. The results of this study will be helpful in future studies to further our knowledge of AD.

References

[1] Avdulov NA, Chochina SV, Igbavboa U, Warden CS, Vassiliev AV, Wood WG. Lipid binding to amyloid β-peptide aggregates: preferential binding of cholesterol as compared with phosphatidylcholine and fatty acids. *J. Neurochem.*69: 1746-1752, 1997.

[2] Bickel PE, Scherer PE, Schnitzer JE, Oh P, Lisanti MP, Lodish HF. Flotillin and epidermal surface antigen define a new famility of caveolae-associated integral membrane proteins. *J. Biol. Chem.* 272: 13793-13802, 1997.

[3] Cenedella RJ, Shi H. Spatial distribution of 3-hydroxy-3-methylglutaryl coenzyme A reductase messenger RNA in the ocular lens: relationship to cholesterologenesis. *J. Lipid Res.* 35: 2232-2240, 1994.

[4] Choo-Smith LP, Surewicz WK. The interaction between Alzheimer amyloid β (1-40) peptide and ganglioside GM1-containing membranes. *FEBS Lett.* 402: 95-98, 1997.

[5] Choo-Smith LP, Garzon-Rodriguez W, Glabe CG, Surewicz WK. Acceleration of amyloid fibril formation by specific binding of Aβ-(1-40) peptide to ganglioside-containing membrane vesicles. *J. Biol. Chem.*272: 22987-22990, 1997.

[6] Corder EH, Saunders AM, Strittmatter WJ, Schmechel DE, Gaskell PC, Small GW, Roses AD, Haines JL, Pericak-Vance MA. Gene dose of apolipoprotein E type 4 allele and the risk of Alzheimer's disease in late onset families. *Science* 261: 921-923, 1993.

[7] Dietschy JM, Turley SD. Cholesterol metabolism in the brain. *Curr. Opin. Lipidol.* 12: 105-112, 2001.

[8] Ehehalt R, Keller P, Haass C, Thiele C, Simons K. Amyloidogenic processing of the Alzheimer β-amyloid precursor protein depends on lipid rafts. *J. Cell Biol.* 160: 113-123, 2003.

[9] Fassbender K, Simons M, Bergmann C, Stroick M, Lutjohann D, Keller P, Runz H, Kuhl S, Bertsch T, von Bergmann K, Hennerici M, Beyreuther K, Hartmann T. Simvastatin strongly reduces levels of Alzheimer's disease β-amyloid peptides Aβ42 and Aβ40 in vitro and in vivo. *Proc. Natl. Acad. Sci. U S A* 98: 5856-5861, 2001.

[10] Frears ER, Stephens DJ, Walters CE, Davies H, Austen BM. The role of cholesterol in the biosynthesis of β-amyloid, *Neuroreport* 10: 1699-1705, 1999.

[11] Goodrum JF. Cholesterol synthesis is down-regulated during regeneration of peripheral nerve. *J. Neurochem.*54: 1709-1715, 1990.

[12] Harper JD, Lansbury PT, Jr. Models of amyloid seeding in Alzheimer's disease and scrapie: mechanistic truths and physiological consequences of the time-dependent solubility of amyloid proteins. *Annu. Rev. Biochem.* 66:385-407.

[13] Harris JR. In vitro fibrillogenesis of the amyloid β1-42 peptide: cholesterol potentiation and aspirin inhibition. *Micron* 33: 609-626, 2002.

[14] Hayashi H, Igbavboa U, Hamanaka H, Kobayashi M, Fujita SC, Wood WG, Yanagisawa K. Cholesterol is increased in the exofacial leaflet of synaptic plasma membranes of human apolipoprotein E4 knock-in mice. *Neuroreport* 13: 383-386, 2002.

[15] Hayashi H, Kimura N, Yamaguchi H, Hasegawa K, Yokoseki T, Shibata M, Yamamoto N, Michikawa M, Yoshikawa Y, Terao K, Matsuzaki K, Lemere CA, Selkoe DJ, Naiki H, Yanagisawa K. A seed for Alzheimer amyloid. *J. Neurosci.*, 24: 4894-4902, 2004.

[16] Igbavboa U, Avdulov NA, Schroeder F, Wood WG. Increasing age alters transbilayer fluidity and cholesterol asymmetry in synaptic plasma membranes of mice. *J. Neurochem.* 66: 1717-1725, 1996.

[17] Ji SR, Wu Y, Sui SF. Cholesterol is an important factor affecting the membrane insertion of β-amyloid peptide (Aβ1-40), which may potentially inhibit the fibril formation. *J. Biol. Chem* .277: 6273-6279, 2002.

[18] Jick H, Zornberg GL, Jick SS, Seshadri S, Drachman DA. Statins and the risk of dementia. *Lancet* 356: 1627-1631, 2000.

[19] Kakio A, Nishimoto SI, Yanagisawa K, Kozutsumi Y, Matsuzaki K. Cholesterol-dependent formation of GM1 ganglioside-bound amyloid-protein, an endogenous seed for Alzheimer amyloid. *J. Biol. Chem* .276: 24985-24990, 2001.

[20] Kakio A, Nishimoto S, Yanagisawa K, Kozutsumi Y, Matsuzaki K. Interactions of amyloid β-protein with various gangliosides in raft-like membranes: importance of GM1 ganglioside-bound form as an endogenous seed for Alzheimer amyloid. *Biochemistry* 41: 7385-7390, 2002.

[21] Kirsch C, Eckert GP, Mueller WE. Cholesterol attenuates the membrane perturbing properties of β-amyloid peptides, *Amyloid* 9: 149-159, 2002.

[22] Kivipelto M, Helkala EL, Hanninen T, Laakso MP, Hallikainen M, Alhainen K, Soininen H, Tuomilehto J, Nissinen A. Midlife vascular risk factors and late-life mild cognitive impairment: A population-based study, *Neurology* 56: 1683-1689, 2001.

[23] Matsuzaki K, Horikiri C. Interactions of amyloid β-peptide (1-40) with ganglioside-containing membranes, *Biochemistry* 38: 4137-4142, 1999.

[24] McLaurin J, Chakrabartty A. Membrane disruption by Alzheimer β-amyloid peptides mediated through specific binding to either phospholipids or gangliosides. Implications for neurotoxicity, *J. Biol. Chem* .271: 26482-26489, 1996.

[25] Mizuno T, Nakata M, Naiki H, Michikawa M, Wang R, Haass C, Yanagisawa K. Cholesterol-dependent generation of a seeding amyloid β-protein in cell culture, *J. Biol. Chem.*274: 15110-15114, 1999.

[26] Molander-Melin M, Blennow K, Bogdanovic N, Dellheden B, Mansson JE, Fredman. Structural membrane alteration in Alzheimer brains found to be associated with

regional disease development; increased density of gangliosides GM1 and GM2 and loss of cholesterol in detergent-resistant membrane domains. *J. Nourochem.* 92: 171-182, 2005.

[27] Naiki H, Gejyo F. Kinetic analysis of amyloid fibril formation. *Methods Enzymol.* 309: 305-318, 1999.

[28] Nilsberth C, Westlind-Danielsson A, Eckman C B, Condron MM, Axelman K, Forsell C, Stenh C, Luthman J, Teplow D.B, Younkin S G, Naslund J, Lannfelt L. The 'Arctic' APP mutation (E693G) causes Alzheimer's disease by enhanced Aß protofibril formation. *Nat. Neurosci.* 4: 887-893, 2001.

[29] Poirier J, Davignon J, Bouthillier D, Kogan S, Bertrand P, Gauthier S. 1993, Apolipoprotein E polymorphism and Alzheimer's disease, *Lancet* 342: 697-699, 1993.

[30] Popplewell PY, Azhar S. Effects of aging on cholesterol content and cholesterol-metabolizing enzymes in the rat adrenal gland, *Endocrinology* 121: 64-73, 1987.

[31] Puglielli, L., Tanzi, R.E., and Kovacs, D.M., 2003, Alzheimer's disease: the cholesterol connection, *Nat. Neurosci.*6: 345-351.

[32] Refolo LM, Malester B, LaFrancois J, Bryant-Thomas T, Wang R, Tint GS, Sambamurti K, Duff K, Pappolla MA. Hypercholesterolemia accelerates the Alzheimer's amyloid pathology in a transgenic mouse model, *Neurobiol. Dis.*7: 321-331, 2000.

[33] Saunders AM, Strittmatter WJ, Schmechel D, George-Hyslop PH, Pericak-Vance MA, Joo SH, Rosi BL, Gusella JF, Crapper-MacLachlan DR, Growden J, Alberts MJ, Hulette C, Crain B, Goldgaber D, Roses AD. Association of apolipoprotein E allele epsilon 4 with late-onset familial and sporadic Alzheimer's disease, *Neurology* 43: 1467-1472, 1993.

[34] Schroeder F, Nemecz G, Wood WG, Joiner C, Morrot G, Ayraut-Jarrier M, Devaux PF. Transmembrane distribution of sterol in the human erythrocyte. *Biochim. Biophys. Acta.* 1066: 183-192, 1991.

[35] Selkoe DJ. Alzheimer's disease: genotypes, phenotypes, and treatments. *Science* 275:630-631, 1997.

[36] Simons K, Ikonen E. Functional rafts in cell membranes, *Nature* 387: 569-572, 1997.

[37] Simons M, Keller P, De Strooper B, Beyreuther K, Dotti CG, Simons K. Cholesterol depletion inhibits the generation of β-amyloid in hippocampal neurons, *Proc. Natl. Acad. Sci. USA* 95: 6460-6464, 1998.

[38] Simons M, Keller P, Dichgans J, Schulz JB. Cholesterol and Alzheimer's disease: is there a link? *Neurology* 57: 1089-1093, 2001.

[39] Stahlberg D, Angelin B, Einarsson K. Age-related changes in the metabolism of cholesterol in rat liver microsomes, *Lipids* 26: 349-352, 1991.

[40] Strittmatter WJ, Weisgraber KH, Huang DY, Dong LM, Salvesen GS, Pericak-Vance M, Schmechel D, Saunders AM, Goldgaber D, Roses AD. Binding of human apolipoprotein E to synthetic amyloid β peptide: isoform-specific effects and implications for late-onset Alzheimer’s disease, *Proc. Natl. Acad. Sci. USA* 90: 8098-8102, 1993.

[41] Telling GC, Scott M, Mastrianni J, Gabizon R, Torchia M, Cohen FE, De Armond SJ, Prusiner SB Prion propagation in mice expressing human and chimeric PrP transgenes implicates the interaction of cellular PrP with another protein. *Cell* 83:79-90, 1995.

[42] Vance JE, Pan D, Campenot RB, Bussiere M, Vance DE. Evidence that the major membrane lipids, except cholesterol, are made in axons of cultured rat sympathetic neurons, *J. Neurochem* .62: 329-337, 1994.

[43] Vey M, Pilkuhn S, Wille H, Nixon R, DeArmond SJ, Smart EJ, Anderson RGW, Taraboulos A, Prusiner SB. Subcellular localization of the cellular and scrapie prion proteins in caveolae-like membrane domains. *Proc. Natl. Acad. Sci. USA.* 93: 14945-14949, 1996.

[44] Wahrle S, Das P, Nyborg AC, McLendon C, Shoji M, Kawarabayashi T, Younkin LH, Younkin SG, Golde TE. Cholesterol-dependent □-secretase activity in buoyant cholesterol-rich membrane microdomains, *Neurobiol. Dis.* 9: 11-23, 2002.

[45] Wolozin B, Kellman W, Ruosseau P, Celesia GG, Siegel G. Decreased prevalence of Alzheimer's disease associated with 3-hydroxy-3-methyglutaryl coenzyme A reductase inhibitors, *Arch. Neurol.*57: 1439-1443, 2000.

[46] Wood WG, Cornwell M, Williamson LS. High performance thin-layer chromatography and densitometry of synaptic plasma membrane lipids. *J. Lipid Res.*30: 775-779, 1989.

[47] Yamamoto N, Hasegawa K, Matsuzaki K, Naiki H, Yanagisawa K. Environment- and mutation-dependent aggregation behavior of Alzheimer amyloid β-protein. *J. Neurochem.* 90: 62-69, 2004a.

[48] Yamamoto N, Igbabvoa U, Shimada Y, Ohno-Iwashita Y, Kobayashi M, Wood WG, Yanagisawa K. Accelerated Aβ aggregation in the presence of GM1-ganglioside-accumulated synaptosomes of aged apoE4-knock-in mouse brain. *FEBS Lett.* 569: 135-139, 2004b.

[49] Yanagisawa K, Odaka A, Suzuki N, Ihara Y. GM1 ganglioside-bound amyloid β-protein (Aβ): a possible form of preamyloid in Alzheimer's disease. *Nat. Med.* 1: 1062-1066, 1995.

[50] Yanagisawa K, McLaurin J, Michikawa M, Chakrabartty A, Ihara Y. Amyloid β-protein (Aβ) associated with lipid molecules: immunoreactivity distinct from that of soluble Aβ. *FEBS Lett.* 420: 43-46, 1997.

[51] Yanagisawa K, Ihara Y. GM1 ganglioside-bound amyloid β-protein in Alzheimer's disease brain. *Neurobiology of aging* 19: 65-67, 1998.

[52] Yip CM, Elton EA, Darabie AA, Morrison MR, McLaurin J. Cholesterol, a modulator of membrane-associated Aβ-fibrillogenesis and neurotoxicity. *J. Mol. Biol.* 311: 723-734, 2001.

In: Glycolipids: New Research
Editor: Daiki Sasaki, pp. 171-184

ISBN: 978-1-60456-216-3

Chapter IX

Structures and Fungicidal Activities of Cellobiose Lipids

***Tatyana V. Kulakovskaya*[a], *Alexander S. Shashkov*[b], *Ekaterina V. Kulakovskaya*[a], *Wladyslav I. Golubev*[a], *Alexander I. Zinin*[b], *Yury E. Tsvetkov*[b], *Alexey A. Grachev*[b] *and Nikolay E. Nifantiev*[b]**

[a]G. K. Skryabin Institute of Biochemistry and Physiology of Microorganisms, Russian Academy of Sciences, Pushchino, Moscow Region, 142290, Russia

[b]N. D. Zelinsky Institute of Organic Chemistry, Russian Academy of Sciences, Moscow, Russia

Abstract

Basidiomycetous yeasts *Cryptococcus humicola* and *Pseudozyma fusiformata* secrete into the culture broth glycolipids which are a 16-(β-cellobiosyloxy)-hexadecanoic acid containing OH-groups at C-2 or C-2 and C-15 of fatty acid aglycon and also carrying O-acyl substituents in cellobiosyl fragment. In the case of *Cr. humicola,* 16-(tetra-O-acetyl-β-cellobiosyloxy)-2,15-dihydrohexadecanoic acid (1) and 16-(tetra-O-acetyl-β-cellobiosyloxy)-2-hydroxyhexadecanoic acid (2) were structurally defined as major products, while *P. fusiformata* secreted mainly 16-[6-O-acetyl-2'-O-(3-hydroxyhexanoyl)-β-cellobiosyloxy)-2,15-dihydroxyhexadecanoic acid (3). These compounds are lethal for many yeast and mycelial fungi. They exhibit higher activity at pH ~4.0 and in the temperature range of 20-30 °C, and had similar fungicidal activities against different yeasts including pathogenic *Cryptococcus* and *Candida* species. The cells of tremellaceous yeast *Filobasidiella neoformans* causing systemic cryptococcosis completely died after 30-min incubation with glycolipids 1-3 at 0.02 mg/ml concentrations. The same effect on ascomycetous yeast, including pathogenic *Candida* species, is achieved at five to eight times higher concentrations of glycolipids. Cellobiose lipid 3 of *P. fusiformata* inhibits the growth of phytopathogenic fungi *Sclerotinia sclerotiorum* and *Phomopsis helianthi* more efficiently than cellobiose lipids **1** and **2** from *Cr. humicola.*

Fully O-deacylated analogues of compounds 1-3, namely 16-(β-cellobiosyloxy)-2,15-dihydroxyhexadecanoic acid (4) and 16-(β-cellobiosyloxy)-2-hydroxyhexadecanoic acid (5), were studied together with a totally synthetic compound, 16-(β-cellobiosyloxy)-hexadecanoic acid (6), to assess the structural features determining their fungicidal activity. Compound 6 without OH-group and compound 5 with only one OH-group in the aglycon do not inhibit the growth of *F. neoformans* and *Saccharomyces cerevisiae,* while compound 4 double-hydroxylated in the fatty acid moiety inhibits the growth of both test cultures but at 5-10 times higher concentrations than compounds 1 and 2 from *Cr. humicola.* Thus, the structures of both the carbohydrate part and fatty acid aglycon moiety influence the fungicidal activity of cellobiose lipids under study. The treatment with cellobiose lipids causes the leakage of ATP and potassium ions from yeast cells. It suggests that the fungicidal activity of cellobiose lipids is due to their ability to enhance the nonspecific permeability of the cytoplasmic membrane. The effective concentrations of cellobiose lipids, which induce ATP release and cell death, were similar for the same test cultures. The results show that extracellular cellobiose lipids may be considered as promising fungicides for agriculture, medicine, and veterinary.

Keywords: *cellobiose lipid, fungicide, yeast, fungi, biosurfactant.*

Introduction

Some yeasts and mycelial fungi produce extracellular glycolipids [1-4]. These compounds contain carbohydrates in combination with fatty acids or hydroxy fatty acids. Glycolipids of various types, such as mannosylerythritols, sophorolipids, and cellobiose lipids, produced by *Candida, Geotrichum, Kurtzmanomyces, Pseudozyma,* and *Ustilago* species are believed to act as biosurfactants and facilitate dissolution of organic hydrophobic compounds in the medium and their consumption by microbial cells [1-3]. The role of sophorolipids [5] and mannosylerythritols [6, 7] as biosurfactants was demonstrated under alkane fermentation by the yeasts *Candida bombicola* and *Pseudozyma antarctica.*

In the recent years, it has been found that the above-mentioned glycolipids possess biological activities, which are of interest for investigations in cellular biology and medical science. For example, sophorolipids induce apoptosis in human liver cancer cells [8], enhance immune response [9], and have spermicidal and antiviral activities [10], whereas mannosylerythritol exhibits the affinity to immunoglobulins [11] and increases gene transfection via membrane fusion [12]. Both sophorolipids and mannosylerythoritols stimulate the growth of neurites [13]. Multiple biological activities and potential applications of extracellular yeast glycolipids are discussed in reviews [2, 3].

Cellobiose lipids have been known as extracellular products of *Ustilago maydis* for a long time [14-16]. However, their function and biological activities were still obscure and the observed antibiotic activity was low [15]. It has also been shown that the yeasts *Cryptococcus humicola* [17] and *Pseudozyma fusiformata* [18] produce non-protein compounds which inhibit the growth of a wide range of both ascomycetous and basidiomycetous yeasts, including medically important species of the genera *Filobasidiella,*

Malassezia, and *Trichosporon.* These activities were exhibited mainly at pH 4-5 and due to secretion of cellobiose lipids of different structures. In the case of *Cr. humicola,* 16-(tetra-O-acetyl-β-cellobiosyloxy)-2-hydroxyhexadecanoic acid was defined as a major product [19]; *P. fusiformata* secreted 16-[6-O-acetyl-2'-O-(3-hydroxyhexanoyl)-β-cellobiosyloxy)-2,15-dihydroxyhexadecanoic acid [20]; *Pseudozyma flocculosa* cellobiose lipid possessed C_8-hydroxy acid and two acetyl groups in the cellobiose residue [21, 22].

Thus, the known antifungal cellobiose lipids produced by yeasts differ in their structures. The goal of our study was to assess the structural features of cellobiose lipids, which determine their fungicidal activity.

Materials and Methods

Strains and growth conditions. Yeasts *Cr. humicola* 9-6 and *P. fusiformata* VKM Y-2821 producing glycolipids were obtained from the Russian Collection of Microorganisms (VKM, http://www.vkm.ru) and grown in a liquid medium as described earlier [23, 24]. Yeast strains *Cryptococcus terreus* VKM Y-2253, *Saccharomyces cerevisiae* VKM Y-1173, *Candida albicans* JCM 1542 (Japanese Collection of Microorganisms), *Filobasidiella neoformans* IGC 3957, *Clavispora lusitaniae* IGC 2705 (Gulbekyan Institute Collection, Portugal), *Candida glabrata* CBS 138, *Candida viswanathii* CBS 4024 (Central Collection of Pure Fungi Cultures, Netherlands), and phytopathogenic fungi *Phomopsis helianthi* and *Sclerotinia sclerotiorum* were used as test cultures. The cultures were grown in the media described earlier [23, 24].

Glycolipid purification. Glycolipids were purified from culture broth after separation from biomass [19, 20, 24]. The concentrations of glycolipids were determined by weighing after methanol evaporation.

Preparation of 2,15-dihydroxy- and 15-hydroxy-16-[4-O-(β-D-glucopyranosyl)-β-D-glucopyranosyloxy)hexadecanoic acids (4 and 5, Figure 1). A purified glycolipid preparation of *Cryptococcus humicola* [23] was evaporated to dryness under vacuum to yield 64 mg of compounds, which were then dissolved in absolute MeOH (2 ml) under argon. A solution of KOH (112 mg, 2 mmol) in absolute MeOH (3 ml) was then added with stirring, and the mixture was agitated for 14 h at 20 °C. Then it was cooled to -40 °C; acetyl chloride (0.67 ml, 9.4 mmol) was added dropwise with stirring, the mixture was allowed to warm to 20 °C, and stirring was continued at this temperature for 8 h. The volatiles were then removed under vacuum, the residue was dissolved in a mixture of pyridine (2 ml) and acetic anhydride (2 ml), and the solution was let stand for 1 day at 20 °C. MeOH (2 ml) was added with cooling in an ice-water bath, so that the temperature would not rise above 40 °C. After standing for 0.5 h at 20 °C, the mixture was evaporated to dryness. Toluene (4×5 ml) was distilled from the residue under reduced pressure. The residue was dissolved in $CHCl_3$ (20 ml), filtered through a Celite pad, and evaporated again. The mixture of peracetylated methyl esters of glycolipids 7 and 8 thus formed was separated by column chromatography on silica gel (gradient elution: benzene → benzene – EtOAc 1:1 v/v, then benzene – EtOAc 1:1) to yield individual compounds 7 (10 mg) and 8 (32 mg). Products 7 and 8 (Figure 2) were subjected to saponification to produce glycolipids 4 and 5 as described below for preparing compound

6. Triethylammonium salts of 4 (7 mg) and 5 (21 mg) were obtained by neutralization with triethylamine. Their structures were confirmed by ^{1}H and ^{13}C NMR spectroscopy (Table 1). ESI-MS (–) data for compound 5: calculated for $[M–H]^{-}$ 611.3; found 611.6.

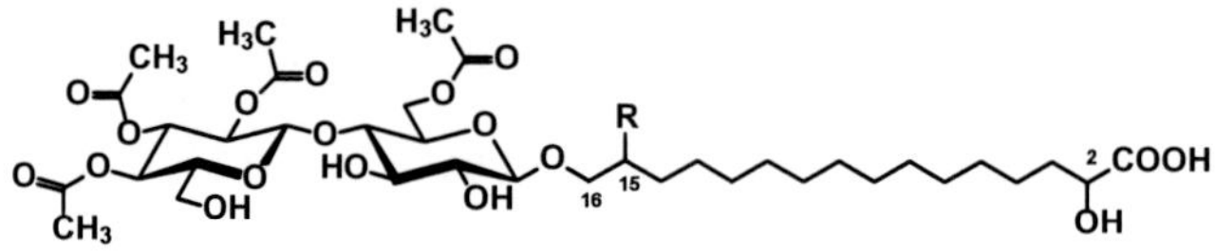

1 R = OH
2 R = H

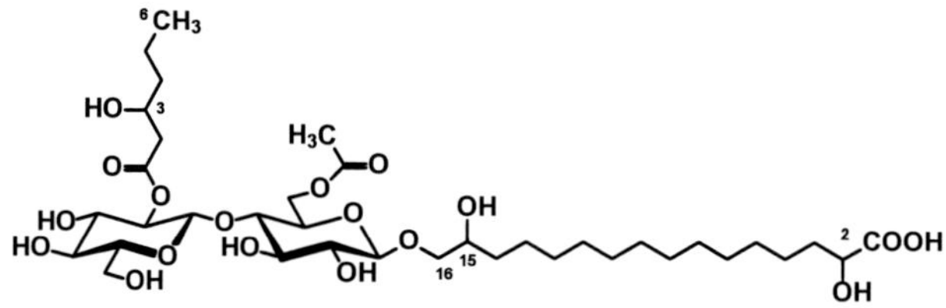

3

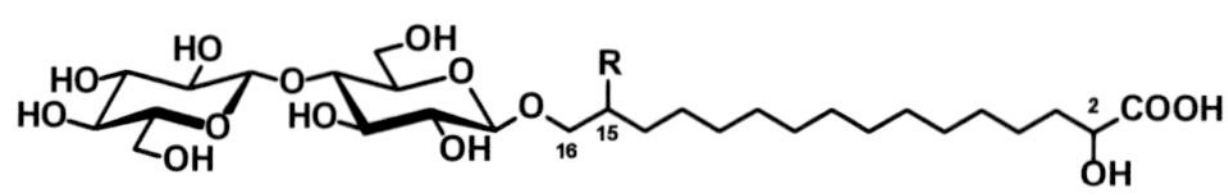

4 R = OH
5 R = H

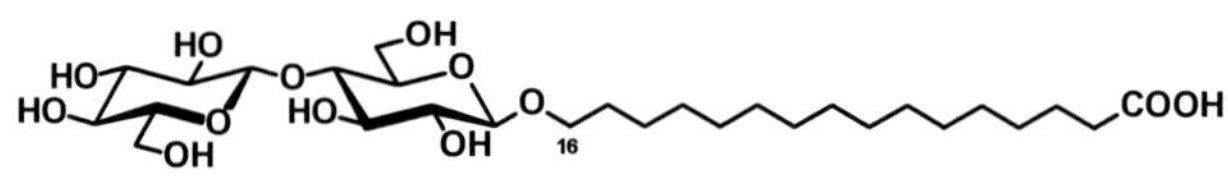

6

Figure 1. The structures of cellobiose lipids from culture broth of Cryptococcus humicola (1,2), *Pseudozyma fusiformata* (3), deacylated derivatives of glycolipids 1 and 2 (4,5) and synthetic compound 16-(β-cellobiosyloxy)-hexadecanoic acid (6).

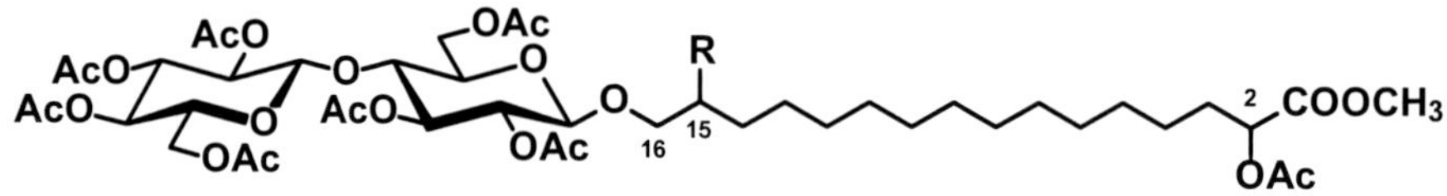

7 R = OAc
8 R = H

Figure 2. The structure of cellobiose lipid 7 and 8.

Table 1. ^{1}H and ^{13}C NMR spectral data of glycolipids in CD_3OD

Structural unit	$J_{1,2}$ or proton number	^{1}H NMR shifts (δ, ppm) and $J_{1,2}$ (Hz)				Carbon number	^{13}C NMR shifts; δ, ppm		
		4	5	6	11**		4	5	6
β-D-Glc-(1→	$J_{1,2}$	7,8	7,9	7,9	n.d.	C-1	104,7	104,6	104,6
	H-1	4,41	4,41	4,42	4,48	C-2	75,0	74,9	74,9
	H-2	3,23	3,22	3,23	4,89	C-3	77,9	77,9	77,9
	H-3	3,37	3,37	3,37	5,11	C-4	71,4	71,4	71,4
	H-4	3,31	3,31	3,30	5,03	C-5	78,1	78,1	78,1
	H-5	3,33	3,34	3,33	3,63	C-6	62,4	62,5	62,4
	H-6	3,88	3,87	3,88	4,34				
	H-6'	3,66	3,65	3,66	4,02				
→4)-β-D-Glc-(1→	$J_{1,2}$	7,8	7,9	7,8	7,9	C-1	104,7	104,2	104,2
	H-1	4,31	4,27	4,27	4,42	C-2	75,0	74,9	74,9
	H-2	3,28	3,23	3,23	4,86	C-3	76,2	76,4	76,5
	H-3	3,52	3,51	3,51	5,15	C-4	80,7	80,8	80,8
	H-4	3,57	3,56	3,56	3,73	C-5	76,5	76,4	76,5
	H-5	3,41	3,39	3,39	3,55	C-6	61,8	61,9	61,9
	H-6	3,88	3,86	3,88	4,47				
	H-6'	3,88	3,86	3,88	4,06				
Aglycon	H-2	3,87	3,87	2.15	2,28	C-1	181,6	181,6	182,7
	H-3	1,74	1,73	1,59	1,60	C-2	73,5	73,5	39,1
	H-3'	1,57	1,55	1,59	1,60	C-3	36,2	36,2	27,7
	H-4	1,43	1,39	1,31	*	C-4	26,4	26,4	30,6
	H-5	*	1,32	*	*	C-5 – C-12	30,7	30,7	30,6
	H-6 – H-12	1,30-1.32	1,29-1,31	1,28 - 1,30	1,22-1,28	C-13	26,6	30,7	30,6
	H-13	1,35	*	*	*	C-14	34,3	27,1	27,1
	H-14	1,43	1,38	1,38	*	C-15	71,7	30,7	30,7
	H-15	3,76	1,62	1,61	1,52	C-16	75,7	71,0	70,9
	H-16	3,89	3,87	3,87	3,80				
	H-16'	3,38	3,53	3,54	3,42				

* This signal is located in the same area with H-6 – H-12 signals.

** Spectrum was recorded in $CDCl_3$, the signal of OCH_3 group was at δ 3,64 ppm.

Preparation of 16-[4-O-(β-D-glucopyranosyl)-β-D-glucopyranosyloxy)] hexadecanoic acid (6, Figure 1). Powdered molecular sieves 4Å (100 mg) was added to a solution of 4-*O*-(2,3,4,6-tetra-*O*-acetyl-β-D-glucopyranosyl)-2,3,6-tri-*O*-acetyl-β-D-glucopyranosyl trichloroacetimidate 9 [25] (80.6 mg, 0.103 mmol) and methyl 16-hydroxyhexadecanoate (10) [26] (36.7 mg, 0.128 mmol, 1.24 eq.) in dry CH_2Cl_2 (1.5 ml) under argon, and the mixture was agitated at 20 °C for 1 h. After cooling to 0 °C, a 0.08 M solution of trimethylsilyl trifluoromethanesulfonate (TMSOTf) in CH_2Cl_2 (0.1 ml) was added dropwise with stirring. More TMSOTf solution was added after 15 min, 60 min, 80 min, and 2 h (0.12 ml, 0.5 ml, 0.75 ml, and 0.9 ml, respectively). The cooling bath was then removed, and the mixture was stirred for 2 h at 20 °C. The reaction was quenched with triethylamine (0.2 ml); the mixture was diluted with benzene, filtered through a Celite pad, and the solvent was removed under vacuum. Column chromatography of the residue (silica gel, gradient elution 10→45% EtOAc/benzene) gave methyl 16-[4-O-(2,3,4,6-tetra-O-acetyl-β-D-glucopyranosyl)-2,3,6-tri-*O*-acetyl-β-D-glucopyranosyloxy]hexadecanoate (11, 24 mg, 26%) as a white foam. The solution of KOH (82 mg) in MeOH (2 ml) was added to the solution of product 11 (47 mg, 52 umol) in MeOH (2 ml) under Ar, and the mixture was stirred for 1 h at room temperature. The resulting mixture was diluted with water (4 ml), and stirring was continued for 16 h at room temperature. Then the mixture was evaporated to dryness in vacuum. The residue was suspended in water (5 ml), acidified with AcOH (0.08 ml), and extracted with water-saturated *n*-BuOH (9 ml and then 2×1 ml). The extracts were combined and evaporated to dryness, and the residue (31 mg, quant.) was crystallized from MeOH – i-PrOH – MeCN to give glycolipid 6 (27 mg, 87%) as fine white needles. Its structure was confirmed by ^1H NMR spectroscopy (Table 1). ESI-MS (–) data for compound 6: calculated for $[M–H]^-$ 595.3; found 595.9. Figure 3 shows the structures of intermediate compounds for the preparation of compound 6.

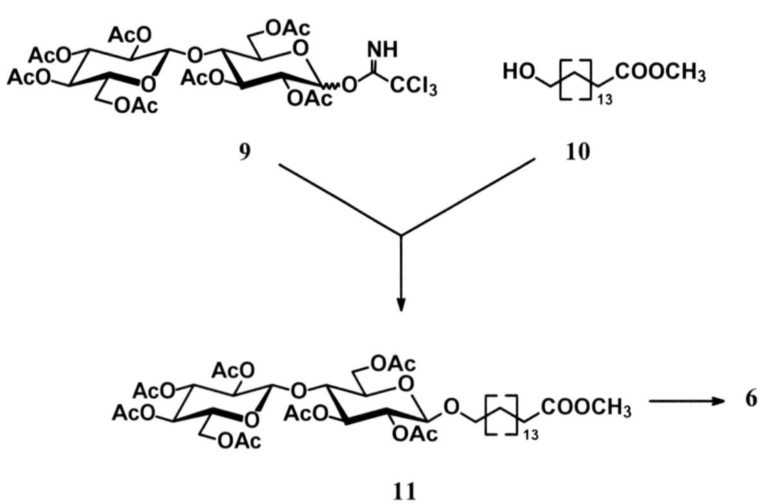

Figure 3. The intermediate compounds for obtaining of 16-[4-O-(β-D-glucopyranosyl)-β-D-glucopyranosyloxy)] hexadecanoic acid (6).

NMR spectroscopy and ESI-MS analysis. One-dimensional ^{1}H-NMR, ^{13}C-NMR, two-dimensional ^{1}H,^{1}H COSY, TOCSY, ROESY, ^{1}H,^{13}C HSQC and HMBC spectra were recorded with a Bruker DRX-500 spectrometer (Germany) at 30 °C. The sample was examined in pyridine-d_5 with TMS as an internal standard. 2D spectra were obtained using standard pulse sequences from the Bruker software. The mixing time of 200 and 100 ms was used in 2D TOCSY and ROESY experiments, respectively. The HMBC experiment was optimized for coupling constant $J_{H,C}$ 5 Hz. The glycolipids 5 and 6 were analyzed by electrospray mass spectrometry (ESI-MS) in the negative mode using a LCO DECAXP mass spectrometer (Thermo Finnigan, USA).

Antifungal activity assay. Cell viability of the test cultures was determined by cultivation on agar medium after incubation with glycolipids in 40 mM citrate-phosphate buffer, pH 4.0, at room temperature for 30 min [23, 24]. The inhibition of growth of the test cultures was also assayed in Petri dishes with glucose-peptone agar containing the same buffer, pH 4.0. Glass fiber disks GF/A (Whatman), 5 mm in diameter, containing different amounts of glycolipids were placed onto the surface of medium inoculated with the test cultures.

The ATP leakage from yeast cells treated with glycolipid preparations was measured by the luciferine-luciferase method [27]. The leakage of potassium ions from the cells of test cultures was measured using a potassium electrode.

Results

The cellobiose lipids secreted by the yeasts *Cr. humicola* and *Ps. fusiformata* differ in their structures as was determined by NMR-spectroscopy (Table 1) and mass-spectrometry. Figure 1 shows the structures of cellobiose lipids used in this study. In the case of *Cr. humicola,* 16-(tetra-O-acetyl-β-cellobiosyloxy)-2,15-dihydroxyhexadecanoic acid (1) and 16-(tetra-O-acetyl-β-cellobiosyloxy)-2-hydroxyhexadecanoic acid (2) were structurally defined as major products in this study, while *Ps. fusiformata* secreted mainly 16-[6-O-acetyl-2'-O-(3-hydroxyhexanoyl)-β-cellobiosyloxy)-2,15-dihydroxyhexadecanoic acid (3) [20].

The cellobiose lipid preparations obtained from the culture broth of *Cr. humicola* and *Ps. fusiformata* demonstrated growth inhibition activities against different yeast species. For example, Figure 4 shows growth inhibition zones under cultivation of test cultures on a solid medium in the presence of different aliquots of cellobiose lipid (3) secreted by *Ps. fusiformata.*

The glycolipid preparation of *Cr. humicola* and compound (3) have almost identical fungicidal activities against yeast species under study (Tables 2, 3). Both cellobiose lipids inhibit the growth of a number of species important for medicine: *C. albicans, C. glabrata, C. viswanathii, F. neoformans,* and *Cl. lusitaniae.* Cellobiose lipids were more efficient against basidiomycetous yeasts of the genera *Cryptococcus* and *Filobasidiella* than against ascomycetous yeasts of the genera *Saccharomyces*, *Candida*, and *Clavispora.*

The growth of mycelial fungi under study was more efficiently inhibited by cellobiose lipid (3) (Table 3). The amount of glycolipid (3) giving growth inhibition zones of 10-20 mm for phytopathogenic fungus *Sclerotinia sclerotiorum* were 5 times lower than that of *Cr. humicola* cellobiose lipids. *Ph. helianthii* and *Mucor mucedo* were less sensitive to the glycolipid from *Cr. humicola.* Even at application of 0.5 mg of this compound onto the disk, the diameters of growth inhibition zones were no more than 8-10 mm.

So, the peculiarities of O-substituents in the cellobiose lipid of *Ps. fusiformata* are important for growth inhibition activity against mycelial fungi.

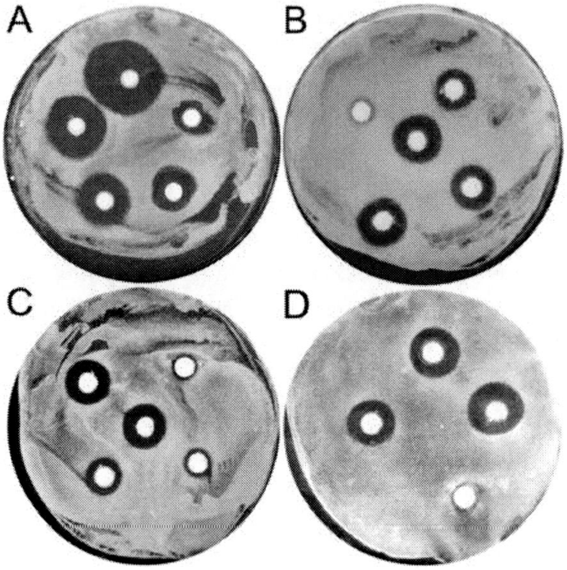

Figure 4. The inhibition of yeast growth by cellobiose lipid (3) of Pseudozyma fusiformata. Test cultures: A - Cryptococcus terreus VKM Y-2253 (0.01, 0.02, 0.03, 0.06, 0.12), B - Filobasidiella neoformans IGC 3957 (0.01, 0.02, 0.03, 0.04, 0.06), C - Candida albicans JCM 1542 (0.03, 0.06, 0.12, 0.2, 0.3), and D - Saccharomyces cerevisiae BKM Y-1173 (0.06, 0.12, 0.2, 0.3). Glucose-peptone agar containing 0.04 M citrate-phosphate buffer, pH 4.0 was used as a solid medium. The amounts of glycolipid (mg/disc) placed on Whatman GF A glass fiber discs are indicated in brackets.

Table 2. The amount of cellobiose lipids of *Cr. humicola* 9-6 and Pseudozyma fusiformata giving 15-20 mm zones of growth inhibition at pH 4.0

Test culture	Glycolipid amount, mg	
	Cellobiose lipid of *Cr. humicola* 9-6	Cellobiose lipid of *Ps. fusiformata*
Cryptococcus terreus	0.05	0.02
Filobasidiella neoformans	0.05	0.02
Clavispora lusitaniae	0.3	0.2
Saccharomyces cerevisiae	0.3	0.2
Candida albicans	0.3	0.2
Candida glabrata	0.3	0.2
Candida viswanathii	0.2	0.1
Sclerotinia sclerotiorum	0.5	0.1
Phomopsis helianthi	-	0.1
Mucor mucedo	-	0.1

Table 3. The concentrations of cellobiose lipids of *Cr. humicola* 9-6 and *Pseudozyma fusiformata* causing almost complete death of the cells of test cultures after 30-min incubation

Test culture	Concentration, mg/ml	
	Cellobiose lipid of *Cr. humicola* 9-6	Cellobiose lipid of *Ps. fusiformata*
Cryptococcus terreus	0.02	0.02
Filobasidiella neoformans	0.02	0.02
Clavispora lusitaniae	0.15	0.1
Candida viswanathii	0.15	0.1
Candida albicans	0.15	0.15
Candida glabrata	0.3	0.3
Saccharomyces cerevisiae	0.3	0.3

The yeast cells treated with the purified glycolipids showed ATP leakage which was maximal at pH 4.0-4.5 and 20-30oC [20, 27]. For example, Figure 5 shows the time dependence of ATP leakage from the cells of Cr. terreus treated by cellobiose lipids of Cr. humicola. The antifungal activities of both compounds were most expressed also at pH 4.5-5.0 and 20oC [17, 18]. The minimal concentrations causing ATP leakage and loss of cell viability were compared for Cr. terreus, S. cerevisiae, and C. albicans. These concentrations were found to be almost equal for the same yeast species (Figure 6 and Table 3). The glycolipids also cause the leakage of K+ ions from the treated cells (Figure 7). Rapid ATP and K+ release from the treated cells of test cultures confirms the previous assumption that the main mechanism of fungicidal action of these compounds is associated with their detergent properties, which provide for the damage of plasma membrane [19]. Cellobiose lipids of *Cr. humicola* and *Ps. fusiformata* showed no differences in the ability to release ATP and K^+ from the cells of the above test cultures. The pH-dependence range of ATP leakage was narrower for cellobiose lipids of *Cr. humicola* as compared with *Ps. fusiformata* (Figure 8). Probably, it is due to the difference in O-substituents in cellobiose residue.

To assess the structural features of cellobiose lipids 1-3, which determine their fungicidal activity, fully O-deacylated analogues of compounds 1-3, namely 16-(β-cellobiosyloxy)-2,15-dihydroxyhexadecanoic acid (4) and 16-(β-cellobiosyloxy)-2-hydroxyhexadecanoic acid (5), were studied together with a totally synthetic compound 16-(β-cellobiosyloxy)-hexadecanoic acid (6). Compounds 4 and 5 were obtained from the glycolipid preparation of *Cr. humicola*. With this purpose, the mixture of glycolipids was subjected to O-deacylation followed by esterification of carboxylic moiety and total O-acetylation. Derivatives 7 and 8 were easily isolated from the mixture of acetylated methyl esters by conventional silica gel column chromatography. Subsequent treatment of 7 and 8 with alkali yielded compounds 4 and 5. Glycosylation of methyl 16-hydroxyhexadecanoate with cellobiosyl trichloroacetimidate 9 gave β-glycoside 10 in moderate yield. Similar yields of β-glycosides were reported at glycosylation of methyl 16-hydroxyhexadecanoate with various glucosyl donors [28]. Saponification of the protected derivative 10 gave the target compound 6.

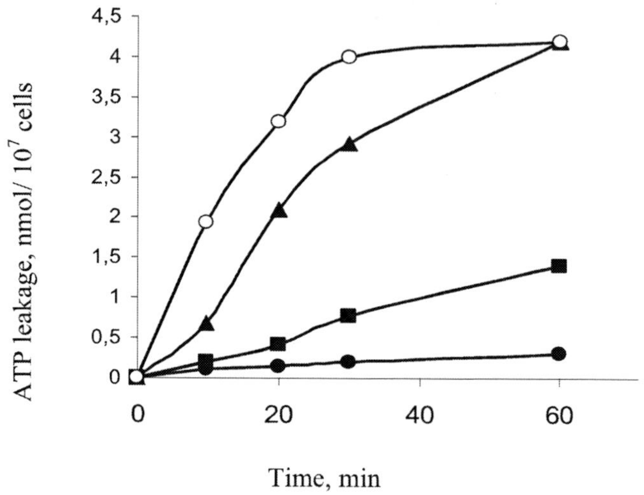

Figure 5. Time dependence of ATP leakage from *Cryptococcus terreus* VKM Y-2253 cells treated with *Cr. humicola* 9-6 glycolipids at pH 4.0 and 20°C. The glycolipid concentrations were ●– 0.013, ■- 0.02, ▲- 0.025 and ○ - 0.04 mg/ml.

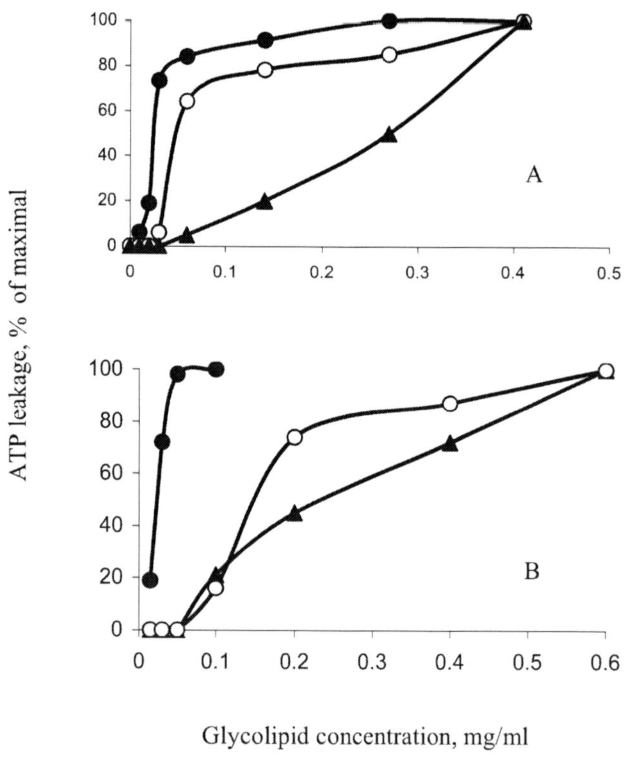

Figure 6. Concentration dependence of ATP leakage from the cells of Cryptococcus terreus VKM Y-2253 (●), Candida albicans JCM 1542 (○), and Saccharomyces cerevisiae VKM Y-1173 (▲) (pH 4,0, 200C) treated by cellobiose lipid (3) - A, and cellobiose lipid of Cryptococcus humicola – B.

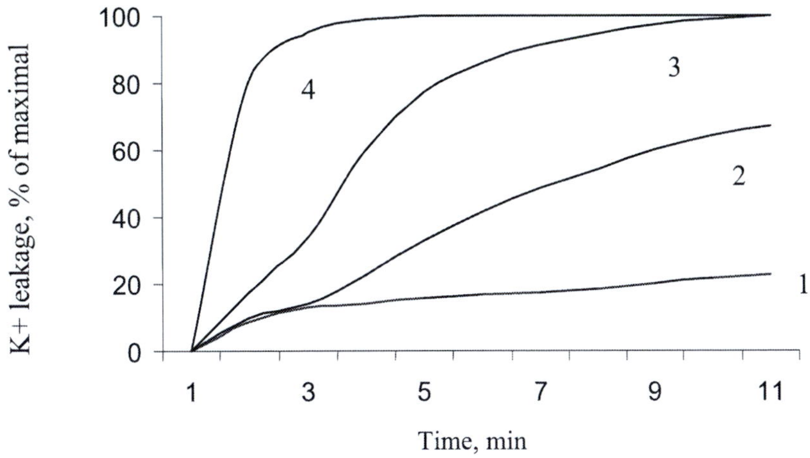

Figure 7. K^+ leakage from *S. cerevisiae* cells treated by cellobiose lipid (3): 1 - 0.14, 2 - 0.22, 3 – 0.29 mg/ml. pH 4.0. K+ leakage under treatment of the cells with 20 mkM Ag^+ (4) is given for comparison.

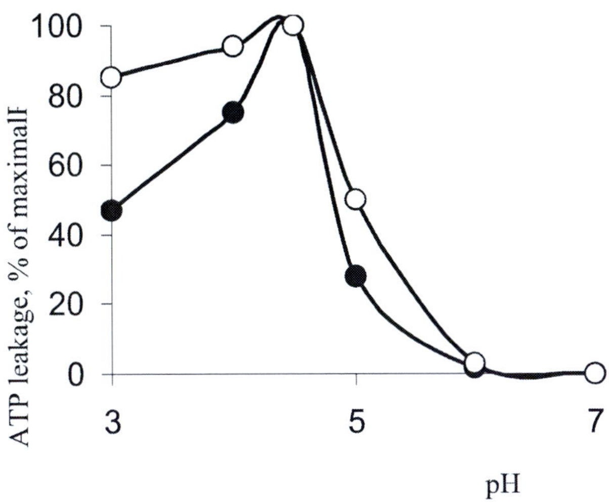

Figure 8. Effects of pH on ATP leakage (for 30 min at 20°C) from cells of *Cryptococcus terreus* VKM Y-2253 treated with glycolipids of Ps. *fusiformata* (o) and *Cr. humicola* (•). The glycolipid concentration was 0.06 mg/ml. No ATP leakage was observed in the control samples with the appropriate methanol concentration.

Compound 6 with no OH-group and compound 5 with only one OH-group in the aglycon do not inhibit the growth of *F. neoformans* and *S. cerevisiae*, while compound 4 double-hydroxylated in the fatty acid moiety inhibits the growth of both test cultures but at 5-10 times higher concentrations than compounds 1 and 2 from *Cr. humicola* (Table 4). Thus, the structures of both the carbohydrate part and fatty acid aglycon moiety influence the fungicidal activity of cellobiose lipids under study.

Table 4. Growth inhibition zones under the action of natural and synthetic cellobiose lipids and their derivates. The amounts of glycolipids per disc are given in brackets

Cellobiose lipid	Test cultures		
	Saccharomyces cerevisiae	*Cryptococcus terreus*	*Filobasidiella neoformans*
(6)	0 (1 mg)	0 (1 mg)	0 (1mg)
(4)	0 (0.8 mg)	0 (0.8 mg) 0 (0.4 mg)	0 (0.8 mg)
(5)	10 mm(0.7mg)	5mm (0.7 mg)	5 (0.7 mg)
Cellobiose lipid of *Cr. humicola*	15 mm (0.3 mg)	15 mm (0.05 mg)	15mm (0.05 mg)

Conclusion

The cellobiose lipids of *Ps. fusiformata* and *Cr. humicola* are highly active against many species of yeast, yeast-like and mycelial fungi (about 300 species of ascomycetous and basidiomycetous organisms) [17, 18]. Their unique structure including OH-groups in fatty acid residue and different O-substituents in cellobiose residue provides fungicidal activity. It should be noted that the effective concentrations of the known fungicides such as nystatine and fluconazoles differ depending on the strains of *Candida albicans* and *Filobasidiella neoformans* (from 0.019 to 10 mg/l) [29]. Taking into account the emergence of strains insensitive to known fungicides, cellobiose lipids are prospective for development of new antifungal compounds. The broad spectra of activity, pH and temperature stability allow one to consider cellobiose lipids as promising natural biocontrol agents of pathogenic yeast and mycelial fungi and to use them in feed and food preservation from fungal spoilage.

Acknowledgments

This work is supported by the Russian Foundation for Basic Research (project 06-04-48215 and 06-04-08253-ofi). We thank A.J. Ivanov for the experiments with K^+ leakage.

References

[1] Spencer, JFT; Spencer, DM; Tulloch, AP. Extracellular glycolipids of yeasts. In: A. H. Rose AH editor. *Economic Microbiology*, vol. 3, Secondary Products of Metabolism, Acad. Press, London, 1979, 523-540.

[2] Kitamoto, D; Isoda, Y; Nakahara, T. Functions and potential applications of glycolipid biosurfactants – from energy-saving material to gene delivery carriers. *J. Biosci. Bioeng.*, 2002, 94, 87-201.

[3] Cameotra, SS; Makkar, RS. Recent applications of biosurfactants as biological and immunological molecules. *Curr. Opin. Microbiol.*, 2004, 7, 262-266.

[4] Golubev, WI. Antagonistic interactions among yeast. In: *Biodiversity and Ecophysiology of the Yeasts,* Berlin Verlag, 2006, 197-219.

[5] Itoh, S; Suzuki, T. Sophorolipids from *Torulopsis bombicola* as microbial surfactants in alkane fermentation. *Appl. Envinron. Microbiol.*, 1982, 43, 1278-1283.

[6] Kitamoto, D; Fujishiro, K; Yanagishita, H; Nakane, T; Nakahara, T. Production of mannosylerythritol lipids as biosurfactants by resting cells of *Candida antarctica. Biotechnol. Lett.*, 1992,14, 305-310.

[7] Kitamoto, D; Ikegami, T; Suzuki, T; Sasaki, A; Takeyama, Y; Idemoto, Y; Koura, N; Yanagishita, H. Microbial conversion of n-alkanes into glycolipid biosurfactants, mannosylerythritol lipids, by *Pseudozyma* (*Candida antarctica*). *Biotechnol. Lett.,*2001, 23, 1709-1714.

[8] Chen, J; Song, X; Zhang, H; Qu, YB; Miao, JY. Sophorolipid produced from the new yeast strain *Wickerhamiella domercqiae* induces apoptosis in H7402 human liver cancer cells. *Appl. Microbiol. Biotechnol.*, 2006, 72, 52-59.

[9] Bluth, MH; Kandil, E; Mueller, CM; Shah, V; Lin, YY; Zhang, H; Dresner, L; Lempert, L; Nowakowski, M; Gross, R; Schulze, R; Zenilman, ME.(2006) Sophorolipids block lethal effects of septic shock in rats in a cecal ligation and puncture model of experimental sepsis. *Crit. Care Med.*, 2006, 34, 188-192.

[10] Shah, V; Doncel, GF; Seyoum, T; Eaton, KM; Zalenskaya, I; Hagver R; Azim, A; Gross, R. Sophorolipids, microbial glycolipids with anti-human immunodeficiency virus and sperm-immobilizing activities. *Antimicrob. Agents Chemother.,* 2005, 49, 4093-4100.

[11] Im, JH; Yanagishita, H; Ikegami, T; Takeyama, Y; Idemoto, Y; Koura, N; Kitamoto, D; Mannosylerythritol lipids, yeast glycolipid biosurfactants, are potential affinity ligand materials for human immunoglobulin G. *J. Biomed. Mater. Res.*, 2003, 65, 379-385.

[12] Inoh, Y; Kitamoto, D; Hirashima, N; Nakanishi, M. Biosurfactant MEL-A dramatically increases gene transfection via membrane fusion. *J. Control. Release*, 2004, 94, 423-431.

[13] Isoda, H; Shinmoto, H; Matsumura, M; Nakahara, T. The neurite initiating effect of microbal extracellular glycolipids in PC12 cells. *Cytotechnology,* 1999, 31, 163-170.

[14] 14 .Lemieux, RU; Thorn, JA; Brice, C; Haskins, RH Biochemistry of the ustilaginales. II. Isolation and partial characterization of ustilagic acid. *Can. J. Chem.*, 1951, 29, 409-414.

[15] Haskins, RH; Thorn, JA Biochemistry of the Ustilaginales. VII. Antibiotic activity of ustilagic acid. *Can. J. Botany*, 1951, 29, 585-592.

[16] Bhattacharjee, SS; Haskins, RH; Gorin, PAJ. Location of acyl groups on two partly acylated glycolipids from strains of *Ustilago* (smut fungi). *Carbohydr. Res.*,1970, 13, 235-246.

[17] Golubev, W; Shabalin, Y. Microcin production in *Cryptococcus humicola. FEMS Microbiol. Lett.*, 2004, 119, 105-110.

[18] Golubev, VI, Kulakovskaia, TV, Golubeva, EV. *Pseudozyma fusiformata* BKM Y-2821-a producer of antifungal glycolipid. *Mikrobiologiia,* 2001, 70, 642-646.

[19] Puchkov, EO; Zahringer, U; Lindner, B; Kulakovskaya, TV; Seydel, U; Wiese, A. Mycocidal, membrane-active complex of *Cryptococcus humicola*, is a new type of

cellobiose lipid with detergent features. *Biochim. Biophys. Acta (Biomembranes)*, 2002, 1558, 161-170.

[20] Kulakovskaya, TV; Shashkov, AS; Kulakovskaya, EV; Golubev, WI. Ustilagig acid secretion by *Pseudozyma fusiformata* strains. *FEMS Yeast Res.*, 2005, 5, 919-923.

[21] Cheng, Y; McNally, DJ; Labbe, C; Voyer, N; Belzyle, F; Belanger, RR. Insertional mutagenesis of a fungal biocontrol agent led to a discovery of a rare cellobiose lipid with antifungal activity. *Appl. Environ. Microbiol.*, 2003, 69, 2595-2602.

[22] Mimee, B; Labbe, C; Pelletier, R; Belanger, RR. Antifungal activity of flocculosin, a novel glycolipid isolated from *Pseudozyma flocculosa*. *Antimicrob. Agents Chemother.*, 2005, 49, 1597-1599.

[23] Kulakovskaya, EV; Golubev, VI; Kulaev IS. Extracellular antifungal glycolipids of *Cryptococcus humicola* yeasts. *Dokl. Biol. Sci.*, 2006, 410, 393-395.

[24] Kulakovskaya, EV; Kulakovskaya, TV; Golubev, WI; Shashkov, AS; Grachev, AA; Nifantiev, NE. Fungicidal activity of cellobiose lipids from culture broth of yeast *Cryptococcus humicola* and *Pseudozyma fusiformata. Russian J. Bioor. Chem.*, 2007, 33, 156-161.

[25] Chang, CWT; Hui, Y; Elchert, B; Wang, J; Li, J; Rai, R. Pyranmycins, a novel class of aminoglycosides with improved acid stability: the SAR of D-pyranoses on ring III of pyranmycin. *Org. Lett.*, 2002, 4, 4603-4606.

[26] Waugh, K.M.; Berlin, K.D. Studies in lipid mimics. Synthesis and carbon-13 relaxation time measurements (T[1] values) of methyl esters of ω-(2-anthryl)alkanoic acids. *J. Org. Chem.*, 1984, 49, 873-878.

[27] Kulakovskaya, TV; Kulakovskaya, EV; Golubev, WI. ATP leakage from yeast cells treated by extracellular glycolipids of *Pseudozyma fusiformata. FEMS Yeast Res.*, 2003, 3, 401-404.

[28] Gouin, SG; Pilgrim, W; Porterb, RK; Murphy, P.V. Synthesis of a glycolipid for studying mechanisms of mitochondrial uncoupling proteins. *Carbohydr. Res.* 2005, 340, 1547–1552.

[29] Carrillo-Munoz, A J; Quindos, G; Tur, C; Ruesga, MT; Miranda, Y; del Valle, O; Cossum, PA; Wallace, TJ. In-vitro antifungal activity of liposomal nystatin in comparison with nystatin, amphotericin B cholesteryl sulphate, liposomal amphotericin B, amphotericin B lipid complex, amphotericin B desoxycholate, fluconazole and itraconazole. *J. Antimicrob. Chemother.*, 1999, 44, 397-401.

Index

A

B

D

F

G

H

I

M

N

O

P

T

U

V

W

Y